Bibliografische Information der Deutschen Nationalbibliothek

Die Deutsche Nationalbibliothek verzeichnet diese Publikation in der
Deutschen Nationalbibliografie; detaillierte bibliografische Daten sind
im Internet über http://dnb.d-nb.de abrufbar.

ISBN 978-3-8325-2848-5

Logos Verlag Berlin GmbH
Comeniushof, Gubener Str. 47,
10243 Berlin
Tel.: +49 (0)30 42 85 10 90
Fax: +49 (0)30 42 85 10 92
INTERNET: http://www.logos-verlag.de

Raman spectroscopy of carboxylic acid and water aggregates

Dissertation

zur Erlangung des Doktorgrades

der Mathematisch-Naturwissenschaftlichen Fakultäten

der Georg-August-Universität zu Göttingen

vorgelegt von

Zhifeng Xue

aus

Qingdao, China

Göttingen 2010

Referent: Prof. Dr. M. A. Suhm
Korreferent: Prof. Dr. B. Abel
Tag der mündlichen Prüfung: 10. Dezember 2010

Zu wissen, was gut ist, ist nicht genug,
man muss es auch lieben.
Die Güte zu lieben, ist nicht genug,
man muss sie auch mit Begeisterung praktizieren.

konfuzius

At this point, I would like to thank all who have contributed through their professional and personal supports to the success of this thesis. Without their help it would have been impossible for me to finish this thesis.

First and foremost, I would like to thank my thesis advisor, Prof. Dr. M. A. Suhm for the interesting and exciting project, his excellent supervision, his outstanding care and the many helpful discussions and talented advices on my work.

Many thanks go to Prof. Dr. B. Abel for the friendly takeover of the co-referee of my thesis.

I would like to thank everyone in the Suhm group for the nice working atmosphere, interesting seminars and the many scientific advices that have always contributed to the improvement of my work. A huge thanks goes to Dr. P. Zielke, who has designed the 1st generation of our experimental setup and helped me continuously and tirelessly in my first year working in the Raman lab. Another huge thanks goes to Dr. T. Wassermann, who has worked with me together after Dr. Zielke in the Raman lab for the improvement of the setup. Both of them have the magic to solve all the big or small problems which occurred in the lab and answer all my countless dumb questions. Never forget Mr. N. Lüttschwager, who will be in charge of the Raman lab in the next several years and has made many progresses of the setup. Take good care of the setup: for all the time it runs well, it is just perfect; for all the "other time", it is also great.

I would like to thank Mr. F. Kollipost, Dr. P. Zielke, Dr. T. Häber, Dr. S. Hesse for their IR spectra used in this work. Special thanks go to Dr. R. Larsen, who measured the very difficult, very controversial and very valuable IR active ν_{24} band of the formic acid dimer with Mr. F. Kollipost for me, using the technically almost "impossible" combination of bolometer and supersonic jet. This band is especially important for me but the only problem is that it was observed just three days before the submission of this thesis.

Thanks to Dr. M. Albrecht for the help on the building and test of the heatable nozzle. For technical help with computer problems, I would like to thank Dr.

U. Schmitt and Dr. I. Dauster. For calculations with Gaussian I am grateful to all group members who have given useful discussions and advice in the group seminars and in the emails. Several students have also participated in this thesis during their undergraduate research. Thanks for their help.

I would like to thank Dr. J. Dreyer from Max-Born-Institut in Berlin to give me the opportunity to work in his group for the spectral simulation of some carboxylic acids and some insight into the computational chemistry.

For the TechnoDudes, I would like to thank all the staffs of the institute workshops and especially Mr. V. Meyer, Mr. A. Knorr and Mr. H. Schlette. Thanks to Mr. W. Noack and Mr. C. Heymann for their constant willingness to help with any kind of technical problem. Besides, Mr. Noack told me many jokes which are not so interesting in my opinion.

I want to extend my thanks to the Graduiertenkolleg 782 as well as the DFG grant Su 121/2 for the financial support to do the research, go to conferences and to present my work. Thanks to the secretaries Ms. P. Lawecki and Mrs. J. Kupferschmid for the kind help with the regular duties.

Personal thanks go to Mr. J. Lee and Mr. N. Lüttschwager for proof-reading my thesis. I know I could make the thesis better and make your proof-reading not so painful, if I was given infinite time. Thanks Prof. Dr. Suhm again to the many good suggestions to improve this thesis and the unbelievable patience to correct the many many misspelled and grammatical mistakes due to my poor English.

I would like to thank all my friends in China and Germany for the many helps and happiness. Your contributions may be invisible but I can feel them everyday.

Finally, I would like to sincerely thank my family who has been always behind me. My parents, my grandmother, my little brother, I miss you everyday. And my dear wife Fei Zhou, thanks for being my sweetheart. You have always supported me with love and patience since the beginning of my studies in Germany in 2003 and suffered so much on my bad temper. I promise life will get easier and better in the future. You will see.

Contents

Contents

Introduction

Microscopic interactions determine the macroscopic processes in all living organisms. Hydrogen bonding is one of the strongest intermolecular interactions. Multiple hydrogen bonds are frequently cooperative in biomolecules, such as proteins and enzymes and enable their conformationally flexible three dimensional structures and functions. Characterization of the hydrogen bonding mechanism is the key to understand these biomolecular functions. Besides, the hydrogen bonding is responsible for many natural phenomena in the environment, for example, building-up of clouds and the solubility of most solvents.

Vibrational spectroscopy (e.g., infrared and Raman spectroscopy) provides a direct access to the local properties of the hydrogen bonding. It can be expected to probe shifts in intramolecular vibrations [1], the newly arising intermolecular vibrations [2–4], the proton tunneling process [5, 6] and the lowest vibrational states of the dimer, which are essential ingredients in a bottom-up statistical-mechanical prediction of the dimerization equilibrium [2, 7].

Long time ago, infrared and Raman investigations in the gas phase at close to room [8,9] or higher temperature [10] have been widely applied for the characterization of the hydrogen bonded systems. However, the thermal excitation results in significant band shape distortions and makes an accurate estimate of the band center difficult unless high resolution spectroscopy reveals the hot band structure [11]. However, the latter is difficult in the Raman case and at low wavenumber in the infrared.

Different kinds of techniques were investigated to reduce the thermal excitation. One method is the application of matrix isolation techniques [12–15]. It gives access to less stable isomers, but may suffer from site splittings due to host-guest interactions and the existence of multiple trapping sites. The spectral simplification offered by this technique comes at the price of matrix shifts, which are sometimes difficult to predict and can lead to changes in resonance patterns [15]. Therefore, it is difficult to obtain a one-to-one correspondence with the gas phase spectrum. Helium nanodroplet spectroscopy is a combination of beam and matrix isolation spectroscopy and has similar characteristics, although the perturbation is significantly smaller.

Another way to get a "cold" spectrum is the supersonic free jet technique. From weakly to strongly bound complexes, combined with IR and Raman, this method has been applied with great success. The development of continuous and pulsed slit jet expansions enables a systematic study of not only clusters of molecules [16] but also the monomers [16]. These cold gas phase spectra can serve as a reference for zero-temperature gas phase calculations [17].

In the present work, spontaneous linear Raman spectroscopy coupled with a supersonic free jet expansion is used to study several simple model system molecules. Water clusters and carboxylic acid dimers are chosen as the prototypes of weakly/strongly bonded systems. The former one is the most important solvent in nature and the exploration of the structural and binding properties of water clusters provides a key for the understanding of many natural phenomena in biological and chemical systems, whereas the carboxylic acid dimers are important for the multiple hydrogen bonded systems. Both of them are highly symmetrical, which means a combination of IR and Raman spectra are necessary. Gaseous water clusters can only be formed at very low temperatures whereas carboxylic acid dimers can be prepared and studied in the room temperature gas phase at high abundance. Therefore, carboxylic acid dimers also represent ideal model systems for the thermodynamic characterization of monomer-dimer equilibria [4, 18, 19]. The double hydrogen bond realized in a carboxylic acid dimer already provides a large fraction of the cohesive energy in the solid or liquid due to the weakness of residual forces when each acid unit is engaged as a single hydrogen bond donor and a single acceptor to another one. Therefore, carboxylic acid dimers can dominate the vapor phase in an almost quantitative fashion, in stark contrast to water dimers which will always strive to continue aggregation, unless they are kinetically hindered to do so.

Beside the water clusters and carboxylic acid dimers, many other molecules or complexes were measured during my doctoral period, for example, neopentyl alcohol [20], water-ethanol complexes [21, 22], $\beta-$Propiolactone [23], β-Butyrolactone [23], γ-Butyrolactone [23,24], methyl lactate [25], chain alkanes [26], acetaldoxime, hydroxyacetone [25] and many complexes of the anesthetics [27] (halothane, sevoflurane, isoflurane, chloroform) with different molecules (dimethyl ether, acetone, benzene). They will not be discussed in detail in this work.

The present work takes the following course: Chap. 1, background information; Chap. 2, technical parameters and characterization of the experimental setup; Chap. 3, monomer and dimer of formic acid; Chap. 4, acetic acid dimer; Chap. 5, pivalic and propiolic acid; Chap. 6, water clusters; Chap. 7. conclusions and outlook of the problems remaining to be solved in the future.

1 Background

Vibrational spectroscopic methods are used for the investigation in this work, mainly Raman spectroscopy (sometimes FTIR), both in the gas phase (GP) and in supersonic expansions. The experimental results are interpreted with the help of quantum chemical calculations. By comparison of the Raman and IR spectra, Davydov splittings of the clusters are analyzed. In the following, some of the most important keywords of Raman spectroscopy, supersonic jet expansion techniques, quantum chemical calculations and Davydov splitting will be introduced.

1.1 Raman spectroscopy

1.1.1 Raman effect

Raman spectroscopy is a spectroscopic technique used to study vibrational, rotational, and other low-frequency modes in gas, liquid or solid phase [28], which relies on Raman scattering. This effect was theoretically predicted by Adolf Smekal in 1923 [29] and experimentally proved by Sir Chandrasekhara Venkata Raman in 1928 [30], who was honored by the Nobel Prize in Physics in 1930.

Systematic pioneering theory of the Raman effect was developed by the Czechoslovak physicist George Placzek between 1930 and 1934 [31]. The Raman effect is interpreted as an inelastic collision between a molecule and a photon. When light impinges upon a molecule, most photons are elastically scattered (Rayleigh scattering) so that the scattered photons have the same frequency ν_0 (energy) as the incident photons. This intense spectral line is called Rayleigh line. A small fraction of the scattered light (approximately 1 in 10 million photons) is scattered by an excitation, with the scattered photons having a frequency different from, and usually lower than the frequency of the incident photons [28] $\nu_0 \pm \nu_k$ (inelastic collision). The frequency difference ν_k between the scattered and incident photon is equal to the difference of the vibrational and rotational energy-levels of the molecule, sometimes also the electronic energy. The vibrational Raman effect is primarily concerned in this work.

Fig. 1.1 shows the mechanism of the two possibilities of Raman scattering in more detail, together with Rayleigh scattering and IR absorption. For the spontaneous Raman effect, a photon excites the molecule from the ground state to a virtual energy state, the energy of which is determined by the incident photon. At the present time, lasers are used as the light source. When the molecule relaxes, it emits a photon and returns to a lower level. If the final state of the molecule is the

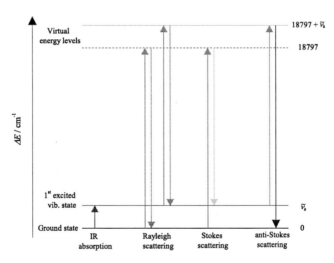

Figure 1.1: Schematic drawings of Raman scattering: Rayleigh, Stokes and anti-Stokes scattering. The virtual energy level at $18797\,\mathrm{cm}^{-1}$ is from the wavenumber of the laser used in this work: $\lambda = 532$ nm. ν_k is the energy difference between two vibrational levels. The IR absorption process is shown for comparison.

same as the original energy level, it is the mentioned elastic Rayleigh scattering. If the final state is more energetic than the initial state, then the emitted photon will be shifted to a lower frequency to conserve the total system energy. This shift in frequency is designated as a Stokes shift. In contrast, if the final state is less energetic than the initial state, then the emitted photon will be shifted to a higher frequency, and this is designated as an anti-Stokes shift. Normally the (vibrational) Stokes line is much stronger than the corresponding anti-Stokes signal due to the population of the molecular energy levels. The intensity ratio of the Stokes and anti-Stokes scattering described in Fig. 1.1 is determined by the Boltzmann distribution according to:

$$\frac{I_{\text{anti-Stokes}}}{I_{\text{Stokes}}} \propto \frac{N_1}{N_0} = \frac{g_1}{g_0} \exp\left(-\frac{h\nu_k}{k_{\text{B}}T}\right). \tag{1.1}$$

where I is the band intensity, N_i is the number of particles with energy E_i, g_i is the degeneracy (the number of states having energy E_i), h is the Planck constant, k_{B} is the Boltzmann constant and T is the temperature.

A change in the molecular polarizability α or amount of deformation of the electron cloud with respect to the vibrational coordinate is required for a vibration to exhibit a Raman effect. When an oscillating electric field vector $\vec{E}$ with the frequency ν_0 impinges upon the molecule, an electric dipole moment $\vec{\mu}_{ind}$ oscillating with the same frequency will be induced in the electron cloud [32], according to the classical electromagnetic theory:

$$\vec{\mu}_{ind} = \alpha\vec{E} = \alpha\vec{E}_0\cos\omega t = \alpha\vec{E}_0\cos 2\pi\nu_0 t. \tag{1.2}$$

During a normal mode vibration the polarizability changes periodically. If the oscillation of the induced dipole moment superimposes that of the molecule, additional components appear in the spectrum of the scattered light. The polarizability can be developed in a linear Taylor series in which the higher order terms are neglected:

$$\alpha = \alpha_{q=0} + (\frac{\partial\alpha}{\partial q})_{q=0} \cdot q + \dots. \tag{1.3}$$

q is the coordinate of the oscillation. The index "$q = 0$" refers to the equilibrium configuration. In the case of a simple harmonic motion, the time dependence of q is given by Eq. 1.4:

$$q = q_0 \cdot \cos 2\pi \nu_k t. \tag{1.4}$$

By a combination of Eqs. 1.2, 1.3 and 1.4, one obtains:

$$
\begin{aligned}
\vec{\mu}_{ind} &= [\alpha_{q=0} + (\frac{\partial \alpha}{\partial q})_{q=0} \cdot q_0 \cdot \cos 2\pi \nu_k t] \cdot \vec{E}_0 \cdot \cos 2\pi \nu_0 t \\
&= \underbrace{\alpha_{q=0} \cdot \vec{E}_0 \cdot \cos 2\pi \nu_0 t}_{\text{Rayleigh scattering}} \\
&\quad + \underbrace{\frac{1}{2} \cdot (\frac{\partial \alpha}{\partial q})_{q=0} \cdot q_0 \cdot \vec{E}_0 \cdot \cos 2\pi (\nu_0 - \nu_k)t}_{\text{Stokes scattering}} \\
&\quad + \underbrace{\frac{1}{2} \cdot (\frac{\partial \alpha}{\partial q})_{q=0} \cdot q_0 \cdot \vec{E}_0 \cdot \cos 2\pi (\nu_0 + \nu_k)t}_{\text{anti-Stokes scattering}}.
\end{aligned}
\tag{1.5}
$$

Eq. 1.5 shows that a vibration is only seen in the Raman spectra when it changes the polarizability of the molecule ($\frac{\partial \alpha}{\partial q} \neq 0$). The frequency of the scattered light is shifted with respect of the incoming light by the vibrational motion frequency ν_k.

The intensity of scattered radiation $I_\perp$ in Raman measurements in the gas phase and in liquid follows the relationship:

$$I_\perp = \sigma' \cdot C \cdot \beta \cdot I_0. \tag{1.6}$$

The index $\perp$ implies the 90° scattering geometry. C is the concentration of molecules, β a proportionality constant, which is dependent on the observation volume of the sample and therefore on the experimental setup (for example, the focal length of the converging lens), and I_0 is the intensity of incoming light. The parameter σ', also written as $\frac{d\sigma}{d\Omega}$, is called scattering cross-section with the unit of $\text{m}^2 \cdot \text{sr}^{-1}$ [33] and describes the ability of vibrational normal modes of molecules to scatter the incoming light inelastically. The scattering cross-section depends on experimental settings like scattering angle and frequency of incoming radiation.

In the following sections the scattering geometry and the scattering cross-section will be briefly explained.

1.1.2 Scattering geometry and depolarization

The scattering intensity is described usually with a general intensity symbol as follows: $I(\theta; P^i, P^s)$. θ is the scattering angle to define the illumination-observation geometry, whereas P^i and P^s represent the states of polarization of the incident (i) and scattered (s) radiation.

In Raman spectroscopy, the samples are routinely measured in the $90°$ scattering geometry or in the $180°$ back scattering geometry. The $90°$ scattering geometry is preferably used for gases and liquids.

This $90°$ scattering geometry is easier to understand in a cartesian, spatially fixed coordinate system (see Fig. 1.2). The direction of propagation of the excita-

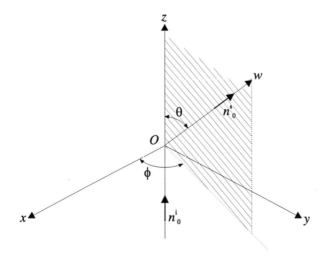

Figure 1.2: Definition of Raman scattering geometry, taken from Ref. [33].

tion laser is on the z axis. The scattering plane is determined by the illumination vector n_0^i and the observation vector n_0^s. Orientations of the vectors to each other are described by the angles θ and ϕ. By rotating the coordinate system about the z-axis by the angle ϕ one can transfer the scattering plane to the xz-plane. Then the angle $\phi = 0$ and only θ must be given. We have $\theta = \pi/2 = 90°$ for the 90°-geometry.

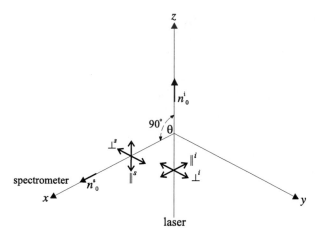

Figure 1.3: Schematic drawing of the 90° Raman scattering geometry.

In Fig. 1.3 the direction of propagation of the excitation laser is on the z axis and the detector on the x axis. The xz plane is the scattering plane, same as in Fig. 1.2. In fact, the incident radiation could be polarized along each axis perpendicular to the z direction. Two limiting cases of P^i are chosen so that the following considerations can be simplified. Incident light with a polarization direction along the y axis is perpendicular to the scattering plane and is therefore characterized with the symbol ($\perp^i$). The other case ($\parallel^i$) is polarization along the x axis. Similarly, the scattered radiation with a polarization perpendicular (along the y axis) or parallel (along the z axis) to the scattering plane is defined as perpendicular ($\perp^s$) or parallel ($\parallel^s$) scattered radiation.

In this work, all the compounds were measured in the gas phase. Compared to the crystal, a gas is a so-called disordered system in which the molecules are oriented statistically without a preferred direction. In a particular direction in space, the observed intensity of the scattered radiation is therefore only a statistical isotropic average [32]. In the case of non-resonant Raman spectroscopy the molecular polarizability α can be assumed to be symmetrical and its frequency dependence can be neglected [34]. However, the spatial direction dependence of α is required for the accurate description of the induced dipole moment $\vec{\mu}_{ind}$. The diagonal and non-diagonal elements obtain the same value after averaging, in the form of a tensor:

$$\alpha = \begin{pmatrix} \langle \alpha_{xx} \rangle & \langle \alpha_{yx} \rangle & \langle \alpha_{zx} \rangle \\ \langle \alpha_{xy} \rangle & \langle \alpha_{yy} \rangle & \langle \alpha_{zy} \rangle \\ \langle \alpha_{xz} \rangle & \langle \alpha_{yz} \rangle & \langle \alpha_{zz} \rangle \end{pmatrix} = \begin{pmatrix} \langle \alpha_{xx} \rangle & \langle \alpha_{xy} \rangle & \langle \alpha_{xy} \rangle \\ \langle \alpha_{xy} \rangle & \langle \alpha_{xx} \rangle & \langle \alpha_{xy} \rangle \\ \langle \alpha_{xy} \rangle & \langle \alpha_{xy} \rangle & \langle \alpha_{xx} \rangle \end{pmatrix}. \tag{1.7}$$

The first sub-index in the tensor specifies which component of the induced dipole is affected, while the second index, from which component of the electric field the effect comes. This nomenclature is also valid for the derivatives of the polarizability along q (see Eq. 1.3). The observed Raman signal consists of the individual tensor components depending on the geometry of excitation, sample and observation.

$\vec{\mu}_{ind}$ in Eq. 1.2 can be described as:

$$\begin{pmatrix} \mu_x \\ \mu_y \\ \mu_z \end{pmatrix} = \begin{pmatrix} \langle \alpha_{xx} \rangle & \langle \alpha_{yx} \rangle & \langle \alpha_{zx} \rangle \\ \langle \alpha_{xy} \rangle & \langle \alpha_{yy} \rangle & \langle \alpha_{zy} \rangle \\ \langle \alpha_{xz} \rangle & \langle \alpha_{yz} \rangle & \langle \alpha_{zz} \rangle \end{pmatrix} \begin{pmatrix} E_x \\ E_y \\ E_z \end{pmatrix}. \tag{1.8}$$

There are two kinds of illumination geometries by reference to the cartesian coordinate system in Fig. 1.3: $\perp^i$ und $\|^i$, following the following relationships, respectively:

$$\vec{E}_{\perp^i}(t) = \begin{pmatrix} 0 \\ E_{y0} \\ 0 \end{pmatrix} \cos \omega t, \tag{1.9}$$

$$\vec{E}_{\parallel^i}(t) = \begin{pmatrix} E_{x_0} \\ 0 \\ 0 \end{pmatrix} \cos \omega t. \qquad (1.10)$$

To calculate the intensity of the light polarized in the y axis ($\perp^s$), only the y-component of the induced dipole moment needs to be considered. The same is valid for the component in the z axis ($\parallel^s$). Combined with the two illumination geometries, there are a total of four possibilities of the states of polarization of the illumination-observation under the 90° scattering geometry: (90°; $\perp^i$, $\perp^s$), (90°; $\perp^i$, $\parallel^s$), (90°; $\parallel^i$, $\perp^s$) and (90°; $\parallel^i$, $\parallel^s$).

The only diagonal (effective) component of the induced dipole moment that can occur with the (90°; $\perp^i$, $\perp^s$) geometry is α_{yy}, and its isotropic average $\langle(\alpha_{yy})^2\rangle$ will be involved in I (90°; $\perp^i$, $\perp^s$):

$$\mu_y = E_x\langle\alpha_{xy}\rangle + E_y\langle\alpha_{yy}\rangle + E_z\langle\alpha_{zy}\rangle = \langle\alpha_{yy}\rangle E_{y_0} \cos \omega t. \qquad (1.11)$$

The intensity of the induced electric field is proportional to the square of the oscillating dipole:

$$I(90°; \perp^i, \perp^s) \propto \langle(\alpha_{yy})^2\rangle. \qquad (1.12)$$

For the (90°; $\perp^i$, $\parallel^s$) geometry:

$$\mu_z = E_x\langle\alpha_{xz}\rangle + E_y\langle\alpha_{yz}\rangle + E_z\langle\alpha_{zz}\rangle = \langle\alpha_{yz}\rangle E_{y_0} \cos \omega t, \qquad (1.13)$$

$$I(90°; \perp^i, \parallel^s) \propto \langle(\alpha_{yz})^2\rangle. \qquad (1.14)$$

For the (90°; $\parallel^i$, $\perp^s$) geometry:

$$\mu_y = E_x\langle\alpha_{xy}\rangle + E_y\langle\alpha_{yy}\rangle + E_z\langle\alpha_{zy}\rangle = \langle\alpha_{xy}\rangle E_{x_0} \cos \omega t, \qquad (1.15)$$

$$I(90°; \parallel^i, \perp^s) \propto \langle(\alpha_{xy})^2\rangle. \qquad (1.16)$$

For the (90°; $\parallel^i$, $\parallel^s$) geometry:

$$\mu_z = E_x\langle\alpha_{xz}\rangle + E_y\langle\alpha_{yz}\rangle + E_z\langle\alpha_{zz}\rangle = \langle\alpha_{xz}\rangle E_{x_0} \cos \omega t, \qquad (1.17)$$

$$I(90°; \|^i, \|^s) \propto \langle(\alpha_{xz})^2\rangle. \tag{1.18}$$

All the isotropic averages of the type $\langle(\alpha_{xx})^2\rangle$ are equal and distinct from those of the type $\langle(\alpha_{xy})^2\rangle$ [32]:

$$\langle(\alpha_{xx})^2\rangle = \langle(\alpha_{yy})^2\rangle = \langle(\alpha_{zz})^2\rangle = \frac{45\alpha^2 + 4\gamma^2}{45}, \tag{1.19}$$

$$\langle(\alpha_{xy})^2\rangle = \langle(\alpha_{yz})^2\rangle = \langle(\alpha_{xz})^2\rangle = \frac{3\gamma^2}{45}. \tag{1.20}$$

Here, α and γ are two parameters to describe the isotropic as well as anisotropic change of the polarizability. In general, the parameters α_k and γ_k are used for the theoretical description of the intensity of the Raman transition in fluids or gases without preferred spatial direction, corresponding to the isotropic and (symmetric) anisotropic modification of the polarizability during the vibration. Both of them are obtained from the isotropic averaging of α derivatives with respect to the spatial coordinate q over all directions (so-called invariants of the tensor) [32]. As mentioned before, the molecular polarizability α can be assumed to be symmetric in the case of non-resonant Raman spectroscopy, and therefore the index k can be neglected:

$$\alpha^2 = \frac{1}{9}\{|\alpha_{xx} + \alpha_{yy} + \alpha_{zz}|^2\}, \tag{1.21}$$

$$\gamma^2 = \frac{1}{2}\{|\alpha_{xx}-\alpha_{yy}|^2+|\alpha_{yy}-\alpha_{zz}|^2+|\alpha_{zz}-\alpha_{xx}|^2\}+3\{|\alpha_{xy}|^2+|\alpha_{yz}|^2+|\alpha_{zx}|^2\}. \tag{1.22}$$

Now we will come back to the intensity relationship of the four combinations of incident linear and scattered linear polarizations described in Eqs. 1.12, 1.14, 1.16 and 1.18. According to Eqs. 1.19 and 1.20, I $(90°; \perp^i, \|^s) = I$ $(90°; \|^i, \perp^s) = I$ $(90°; \|^i, \|^s) \neq I$ $(90°; \perp^i, \perp^s)$. It can be seen from Fig. 1.3 that for I $(90°; \perp^i, \perp^s)$ the electric vectors of the incident and scattered radiation are parallel to each other, whereas for the other three intensities they are perpendicular to each other. This distinction does not involve any axis system, as is to be expected for scattering systems that are isotropically averaged [32].

It follows that only two appropriately chosen intensity measurements involving linear polarized radiation are needed to characterize a symmetric transition polarizability tensor in the cartesian basis. One must be I (90°; $\perp^i$, $\perp^s$) because it alone involves an isotropic average of the type α_{zz}, the other can be any of the three involving isotropic averages of the type α_{xy}, and in this work I (90°; $\|^i$, $\perp^s$) is chosen. The depolarization ratio δ, which is the intensity ratio between the perpendicular component and the parallel component of the Raman scattered light, can be so summarized according to Eqs. 1.12, 1.16, 1.19 and 1.20:

$$\delta = \frac{I(90°; \|^i, \perp^s)}{I(90°; \perp^i, \perp^s)} = \frac{3\gamma^2}{45\alpha^2 + 4\gamma^2} \leqslant 0.75. \tag{1.23}$$

$0 \leq \delta \leq 0.75$ because γ^2 and α^2 are non-negative. A totally symmetric vibration produces scattered light, whose original polarization is preferably maintained. Raman lines with $\delta = 0$ are described as completely polarized, which only arise from the totally symmetric vibrations of optically isotropic molecules with $\gamma = 0$. The Raman lines of all not totally symmetric vibrational modes are completely depolarized with $\alpha = 0$ and produce $\delta = 0.75$. Totally symmetric vibrations of optically anisotropic molecules produce $0 < \delta < 0.75$.

δ can be experimentally examined by using a polarizing filter. Light is permeable polarized separately parallel ($\|^s$) or perpendicular to the scattering plane ($\perp^s$) and the various contributions to total intensities can be measured.

Since the use of a polarization filter brings a loss of light intensity except at $\delta = 0$, a more efficient solution to the polarization measurement may occur through the use of a so-called $\lambda/2$ retardation plate. Two measurements are carried out by rotating the excitation laser from its perpendicular polarization relative to the scattering plane ($\perp^i$) into a parallel orientation ($\|^i$). In this case, both the $\|^s$- and the $\perp^s$-component of the scattered light contribute to the signal intensities and a depolarization ratio δ' can be then re-determined:

$$\delta' = \frac{I(90°; \|^i, \perp^s + \|^s)}{I(90°; \perp^i, \perp^s + \|^s)} = \frac{3\gamma^2 + 3\gamma^2}{45\alpha^2 + 4\gamma^2 + 3\gamma^2} = \frac{6\gamma^2}{45\alpha^2 + 7\gamma^2} \leqslant 0.857. \tag{1.24}$$

The values of δ can be obtained with quantum chemical calculations and thus compared with the experimental data. The depolarization measurements can therefore assign the bands in the spectrum under investigation: bands which are reduced in intensity by more than a factor of 6/7 are due to totally symmetric vibrations ($\alpha \neq 0$), according to Eq. 1.24.

1.1.3 Scattering cross-section

Eq. 1.6 shows that the relative intensities of different bands in the same spectrum depend only on the scattering cross-section σ' ($\frac{d\sigma}{d\Omega}$). For the $(90°; \perp^i, \perp^s + \|^s)$ geometry, σ' can be calculated as [32]:

$$\sigma'\left(90°; \perp^i, \perp^s + \|^s\right) = \frac{2\pi^2 h}{45c\tilde{\nu}_k} \cdot \frac{(\tilde{\nu}_0 - \tilde{\nu}_k)^4}{1 - \exp\left(-\frac{hc\tilde{\nu}_k}{k_BT}\right)} \cdot g_k A_s. \quad (1.25)$$

c is the speed of light, $\tilde{\nu}_0$ and $\tilde{\nu}_k$ are the wavenumbers of irradiated light and of the molecular vibrations with q_k as normal coordinate (g_k: degeneracy factor), and A_s is the scattering activity, which can be obtained from the output of the GAUSSIAN 03 [35] software:

$$A_s = 45\alpha_k^2 + 7\gamma_k^2, \qquad [A_s] = \text{Å}^4/\text{amu}. \quad (1.26)$$

To compare theoretical and experimental spectra A_s has to be transformed into the scattering cross-section, as Eq. 1.25 shows.

The factor $\left(1 - \exp\left(-\frac{hc\tilde{\nu}_k}{k_BT}\right)\right)$ comes from the summation over all excited levels at a given temperature T [31]. It assumes the same frequency for the "hot bands" of a harmonic oscillator, but is not valid in the anharmonic experimental case, where the positions of the hot bands may move relative to that of the fundamental transition.

The effective scattering cross-section increases by the thermal excitation to high excited levels, as the higher level vibrations scatter more strongly due to the larger nuclear amplitudes [31]. This effect is noticeable for the low frequency vibrations, whose excited levels are more easily occupied. On the other hand, the low frequency vibrations are more effectively cooled by the collisions in a jet expansion.

In this work, the factor $\left(1 - \exp\left(-\frac{hc\tilde{\nu}_k}{k_\mathrm{B}T}\right)\right)$ is set to 1 (equivalent to $hc\tilde{\nu}_k \gg k_\mathrm{B}T$), because the accurate temperatures of the individual vibrational degrees of freedom in the expansion are unknown. Under these conditions, a proportionality of $1/\tilde{\nu}_k$ remains as the dominant contribution to the band intensity for the low wavenumbers $\tilde{\nu}_k$ (i.e. at frequencies of scattered light near to the Rayleigh line).

Eq. 1.25 builds on Eq. 1.6, in which the intensities I of the scattered light as well as the incident laser light refer to an "energy density". It works by measuring the energy of the scattered light directly [36]. In contrast, if the signal is measured with a CCD camera as in this work, which means the detection is carried by counting photons, the "photon density" Φ should be used instead of the "energy density" as unit of the signal intensities. The energy should be divided by $hc\tilde{\nu}$ ($E = h\nu = hc\tilde{\nu}$) to be converted into a photon density, where for $\tilde{\nu}$ the wavenumbers of the scattered light $\tilde{\nu}_s = (\tilde{\nu}_0 - \tilde{\nu}_k)$ and of the laser light $\tilde{\nu}_0$ should be used:

$$\Phi_\perp = \frac{I_\perp}{hc(\tilde{\nu}_0 - \tilde{\nu}_k)} = \sigma' \cdot C \cdot \alpha \cdot \frac{I_0}{hc\tilde{\nu}_0} \cdot \frac{hc\tilde{\nu}_0}{hc(\tilde{\nu}_0 - \tilde{\nu}_k)} = \sigma' \cdot C \cdot \alpha \cdot \Phi_0 \cdot \frac{\tilde{\nu}_0}{(\tilde{\nu}_0 - \tilde{\nu}_k)}. \quad (1.27)$$

The differential scattering cross-section σ'_Φ for the case of photon counting can thus be obtained from the differential scattering cross section σ' for the energy density measurement:

$$\sigma'_\Phi = \sigma' \cdot \left(\frac{\tilde{\nu}_0}{\tilde{\nu}_0 - \tilde{\nu}_k}\right). \quad (1.28)$$

The scattering cross-section in Eq. 1.25 can be transformed directly as follows:

$$\sigma'_\Phi\left(90°; \perp^i, \perp^s + \|^s\right) = \frac{2\pi^2 h}{45c\tilde{\nu}_k} \cdot \frac{\left(\tilde{\nu}_0 - \tilde{\nu}_k\right)^3 \tilde{\nu}_0}{1 - \exp\left(-\frac{hc\tilde{\nu}_k}{k_\mathrm{B}T}\right)} \cdot g_k A_\mathrm{s}. \quad (1.29)$$

The value of correction in the Eq. 1.29 depends on the wavenumber $\tilde{\nu}_k$ for constant wavenumber of the incident light $\tilde{\nu}_0$. $\left(\frac{\sigma'_\Phi}{\sigma'}\right)$ is $\geqslant 1$ for $\tilde{\nu}_k < \tilde{\nu}_0$. Both the deviation from linear behavior and the amount of the correction factor grow with wavenumber.

In this work, σ'_Φ will be always calculated as the scattering cross-section due to the detection way of photo counting. The calculated values will be directly given as $\frac{d\sigma}{d\Omega}$ and the index Φ will be neglected.

The scattering cross-section ratio of two bands in the same Raman spectrum can be calculated as:

$$\frac{\Phi_{\tilde{\nu}_1}}{\Phi_{\tilde{\nu}_2}} = \left(\frac{\tilde{\nu}_0 - \tilde{\nu}_1}{\tilde{\nu}_0 - \tilde{\nu}_2}\right)^3 \cdot \left[\frac{1 - \exp(-hc\tilde{\nu}_2/k_BT)}{1 - \exp(-hc\tilde{\nu}_1/k_BT)}\right] \cdot \frac{A_s^1 \tilde{\nu}_2 g_1}{A_s^2 \tilde{\nu}_1 g_2}. \tag{1.30}$$

If the two compared bands are from different molecules or from different molecular conformers with inequivalent energies, the ratio of the occupation numbers of the two molecules/conformers for the Boltzmann distribution $\frac{N_1}{N_2}$ should be introduced in Eq. 1.30:

$$\frac{N_1}{N_2} = \exp\left(\frac{E_2 - E_1}{RT_k}\right). \tag{1.31}$$

The energy difference $(E_2 - E_1)$ can be calculated or experimentally determined. The above listed equations were all derived for Stokes scattering, following:

$$\tilde{\nu}_s = \tilde{\nu}_0 - \tilde{\nu}_k. \tag{1.32}$$

The related scattering cross-section of anti-Stokes scattering can be calculated analogously:

$$\tilde{\nu}_s = \tilde{\nu}_0 + \tilde{\nu}_k. \tag{1.33}$$

The intensity ratio between Stokes and anti-Stokes scattering $\frac{I_S}{I_{AS}}$ and $\frac{\Phi_S}{\Phi_{AS}}$ can be calculated by transformation of Eq. 1.30, with the same Raman activities:

$$\frac{I_S}{I_{AS}} = \frac{(\tilde{\nu}_0 - \tilde{\nu}_k)^4}{(\tilde{\nu}_0 + \tilde{\nu}_k)^4} \cdot \exp\left(\frac{hc\tilde{\nu}_k}{k_BT}\right), \tag{1.34}$$

$$\frac{\Phi_S}{\Phi_{AS}} = \frac{(\tilde{\nu}_0 - \tilde{\nu}_k)^3}{(\tilde{\nu}_0 + \tilde{\nu}_k)^3} \cdot \exp\left(\frac{hc\tilde{\nu}_k}{k_BT}\right). \tag{1.35}$$

These relationships can be used for the determination of the (vibrational) temperatures.

1.2 Supersonic jet expansion

The supersonic free-jet expansion technique is a powerful method to simplify rotational-vibrational spectra and has been used in different fields of research in physics, physical chemistry and chemistry. In the present work, the Raman/IR spectroscopy coupled with the free jet expansion serves as the investigation method to study molecular clusters in a unique environment. With the significant reduction of the thermal effect and much simplified band structures in comparison with the GP spectra at room temperature, we are able to assign the complicated band structures of the carboxylic dimers and the weak, broad and overlapping bands of water clusters.

A detailed explanation of free jet expansions can be found in Ref. 37. A gas or gas mixture is expanded adiabatically from a reservoir at moderate pressure p_S through a pinhole or slit nozzle into a vacuum chamber at reduced pressure p_∞, leading to low rotational temperatures and to some vibrational cooling of the expanded molecules. In the experiments in this work, typical conditions were $p_S = 0.2\text{-}0.8\,\text{bar}$, $p_\infty = 1\text{-}3\,\text{mbar}$.

The ratio of the clusters with different sizes depends strongly on the number and efficiency of collisions in the jet. The population of energetically higher lying states may be drastically reduced in the expansion so that a cooling of the internal degrees of freedom results. The cooling efficiency differs for the different degrees of freedom. Translational and rotational energies can be efficiently transferred by a few collisions.

In our Raman measurements we vary the distances d from laser to nozzle exit to regulate the number of collisions. At the nozzle opening, the Mach number M, the ratio between the flow rate u and the local sound-propagation velocity a is one. It then increases until the molecules reach the Mach disk at a distance d_M from the nozzle where they are slowed down to thermal velocities by the incoming shock waves. The distance from the nozzle to the Mach disk d_M is dependent on p_0, p_∞ and the nozzle geometry.

For a pinhole nozzle, the dependence is given by [38]:

$$d_M = 0.67\sqrt{\frac{p_0}{p_\infty}} \cdot d' \tag{1.36}$$

and for a slit nozzle, the equation is given by:

$$d_M = D \cdot 0.67\sqrt{\frac{p_0}{p_\infty}} \cdot \left(\frac{L}{D}\right)^\varepsilon. \tag{1.37}$$

L is the length of the slit, D is the slit width, ε is an empirical parameter, which is between 0.47 and 0.74, and d' is the diameter of the pinhole nozzle. This only holds true when the L/D is smaller than 50.

The efficiency of the cooling depends also on the kind of the carrier gas. In this work, different noble gases (helium, neon and argon) as well as noble gas mixtures were used. Atomic gases only possess translational degrees of freedom, therefore they reach very low temperatures. Furthermore, noble gases have no Raman activity as long as they do not cluster. Argon leads to better jet-cooling because it is a more efficient collision partner; however, its dispersion and induction forces ultimately lead to condensation on the molecules.

There is no strictly defined thermodynamic equilibrium in the zone of silence in a jet. Nevertheless, the rotational as well as vibrational temperatures can be determined in different ways, all based on the Boltzmann contribution. Detailed descriptions of the methods can be found in Ref. 39.

1.3 Quantum chemical calculations

In the present work, quantum chemical calculations were carried out using the Gaussian 03 and Gaussian 09 program packages [35,40] to support and interpret the experimental assignments. Optimized structures, harmonic frequencies, the relative band intensities and some thermodynamic data were calculated.

Before a structure can be optimized or the harmonic frequencies calculated, a method and basis set must be chosen. Methods, which were normally used during this work, include: Hartree Fock (HF) theory, hybrid Density Functional Theory (DFT) methods (B3LYP) and Møller-Plesset second order perturbation theory (MP2). Many different kinds of basis sets are used. Quantum chemical calculations are described in detail in Ref. [41]. Two approximations are made in all methods exploited in this work, relativistic effects are neglected and the electron and nucleus movement was separated (Born-Oppenheimer approximation).

HF is the simplest of all the methods and includes no electron correlation. It is a self-consistent field (SCF) calculation method. DFT exploits the fact that the ground state energy of a molecule is a functional of its electron density. The electronic energy of DFT functionals is split into the exact terms describing the kinetic energy, nucleus-electron-attraction, the Coulomb energy between electrons and the residual, unknown terms of exchange and correlation energy. The B3LYP functional was used in this work. It is a functional combination, using the B3 exchange functional by Becke and the LYP correlation functional by Lee, Yang and Parr. MP2 includes electron correlation on top of HF calculations in a perturbative way. This gives a better description of intermolecular interactions but is computationally more demanding. For the comparison of the low frequency fundamentals of several carboxylic acids (see Chap. 5.1.1), a dispersion-corrected density functional (B97D) implemented in the Gaussian 09 program package [40] was used.

A suitable basis set should be selected to be used with the methods. It provides a mathematical description of the orbitals. Usually, Gaussian type orbitals (GTOs) are used as basis functions instead of the more physically motivated Slater type orbitals (STOs). Due to the variational principle, the accuracy depends on the size of the basis set, which is the linear combinations of several Gaussian functions such as 3-21+G*, 6-31+G and 6-311+G*. The basis set 6-311+G* will be explained as an example. This basic set includes six contracted Gaussian functions for the inner shell orbitals and valence orbitals which are composed of five primitive Gaussian functions. Three of them are optimized together in a contracted Gaussian function, while the other two functions are optimized independently. A + denotes the addition of extra diffuse functions whereas a * represents additional basis functions with higher angular momentum quantum number, for example, d orbitals for oxygen.

Computational costs rise steeply with the number of basis functions. Therefore, making a basis set as small as possible without sacrificing accuracy is of prime importance. In this work, comparisons between the experimental results with the quantum chemical calculations at several different levels using various methods are always made.

Anharmonic force field calculations are carried out to assess the influence of anharmonicity on the fundamentals and to give out the band positions of the overtone/combination bands, which is especially useful for the band assignments of the carboxylic acid dimers [42]. Several calculation results with counterpoise (CP) correction [43] were also analyzed and comparisons with our calculations were made. It is the most popular means of correcting the basis set superposition error (BSSE) of a system containing more units. The CP method calculates each of the units including the basis functions of the other (without the nuclei or electrons), using so-called "ghost orbitals".

1.4 Davydov splitting

The two main model systems studied in this work are carboxylic acid dimers and water clusters. Both are regarded to be with (local) rotational symmetry and equivalent oscillators (see Fig. 1.4). Due to the inversion center most of their vibrations occur in Davydov pairs of IR active and Raman active modes. The Davydov splitting, also named factor-group splitting, is defined by the IUPAC as the splitting of bands in the electronic or vibrational spectra of crystals due to the presence of more than one (interacting) equivalent molecular entity in the unit cell [44]. In this work, it is taken as the group of IR and Raman active vibrational bands which arise from a single vibration motion of the monomer due to interaction in the cluster.

The degeneracy of the local oscillators is lifted by the intermolecular coupling, and the mutually orthogonal linear combinations are energetically lowered or

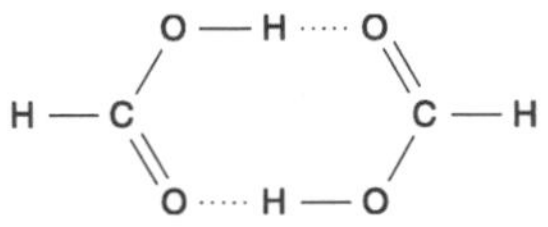

Formic acid dimer 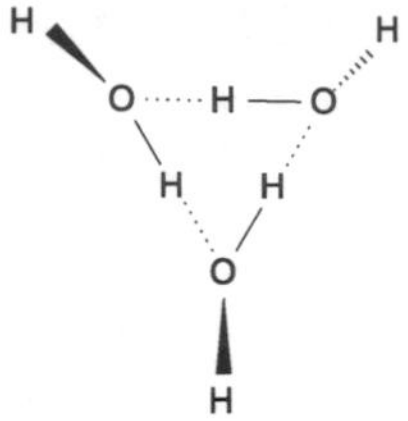   Water trimer

Figure 1.4: Two systems studied in this work, which are with (local) rotational symmetry and equivalent oscillators. The formic acid dimer is a C_{2h}-symmetric system whereas the water trimer shows a pseudo-C_3-symmetric system.

raised. Since the oscillators are located on separate molecules, the coupling in the cooperative hydrogen bonded systems is (unlike for example the stretching vibrations of CO_2) considered as pure potential coupling. Usually, the totally symmetric linear combination lies on the lowest energy level. It shows only a very small (e.g., C_n symmetry) or even no (C_i, C_{nh}, S_n symmetry, etc.) change of the dipole moment. However, the change in the polarizability and the associated Raman activity is relatively strong. For the systematic investigation of the Davydov splitting the combination of absorption (e.g., IR) and Raman spectroscopy is required.

In order to compare the couplings of different cluster sizes and geometries, it is useful to determine the coupling matrix elements similar to the Hückel-MO-theory. For a cyclic dimer (e.g., carboxylic acid dimer) the coupling determinant is:

$$\begin{vmatrix} X & W \\ W & X \end{vmatrix}.$$

W is the coupling matrix element. The determinant is zero for

$$X = \pm W,$$

the eigenvalues of the corresponding matrix. The Davydov splitting is thus $2W$.

Considering the case of the symmetric, cyclic trimer, each oscillator can couple with two other neighbor elements. Since the unperturbed oscillators are equivalent, the coupling matrix elements are same for each coupling. The determinant equation to be solved is as follows:

$$\begin{vmatrix} X & W & W \\ W & X & W \\ W & W & X \end{vmatrix} = 0.$$

The Davydov splitting between the A and E symmetric modes is $3W$.

For the treatment of the symmetric, cyclic tetramers, a further coupling matrix element must be introduced. Each oscillator has two direct neighbors and one indirect one. The coupling with the direct neighbors is described with the matrix

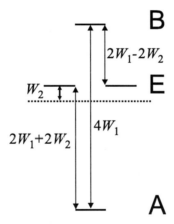

Figure 1.5: Energy diagram of the Davydov splitting of the symmetric, cyclic tetramers in dependence of the coupling matrix elements W_1 und W_2, taken from Ref. 33.

element W_1. W_2 describes the coupling with the second closest, opposite oscillator. The determinant equation is as follows:

$$\begin{vmatrix} X & W_1 & W_2 & W_1 \\ W_1 & X & W_1 & W_2 \\ W_2 & W_1 & X & W_1 \\ W_1 & W_2 & W_1 & X \end{vmatrix} = 0.$$

The A/B splitting is therefore $4W_1$, the E/B splitting is $2W_1 - 2W_2$ and the A/E splitting is $2W_1 + 2W_2$, as the energy level scheme clarifies in Fig. 1.5.

In the case of the symmetric, cyclic pentamers, the coupling matrix is more complicated to be solved and will not be discussed here. Descriptions of the coupling systems and the related Davydov splittings of the cyclic-distorted system with "broken" symmetries can be found in Ref. 33.

2 Experimental

2.1 *Curry*-jet setup

In this thesis a FTIR-jet spectrometer named *filet*-jet and a spontaneous Raman scattering jet spectrometer named *curry*-jet were used for the measurements. The *filet*-jet, which stands for the *fine*-but-*length*y nozzle, has already been described in detail in Refs. [5, 25, 37, 45–48] and will be not further introduced in this work. The acronym *curry*-jet stands for <u>c</u>lassical <u>unr</u>estricted <u>R</u>aman spectroscop<u>y</u> in a jet. It is a modified version of the Raman-jet spectrometer built up by P. Zielke during his doctoral thesis [33, 49, 50]. Changes and enhancements became possible through the support of a DFG grant [51]. A detailed comparison between the two generations of setup is provided in Ref. 39. Most measurements in this thesis were carried out with the new setup.

The new setup is illustrated in Fig. 2.1. The operating principle has been described previously [4, 39, 52–54] and is schematically shown in Fig 2.2. The carrier gas (normally pure helium) flows through a thermostatted glass saturator which contains the studied compounds. The resulting compound concentration in the gas mixture is estimated by estimating its equilibrium vapor pressure and assuming near-saturation of the carrier gas (normally 1.5~1.6 bar) so that the concentration can be controlled by the temperature of the saturator T_s as well as the carrier gas pressure p_0. For measurements of high vapor pressure substances, gas mixtures are prepared and stored in a separate gas admixture bottle [27]. The gas mixture is collected in a 67 L stainless steel reservoir (until early 2009) or a 4.7 L Teflon-coated reservoir and expanded through a homebuilt slit nozzle with different geometrical parameters into an aluminum chamber with dimensions

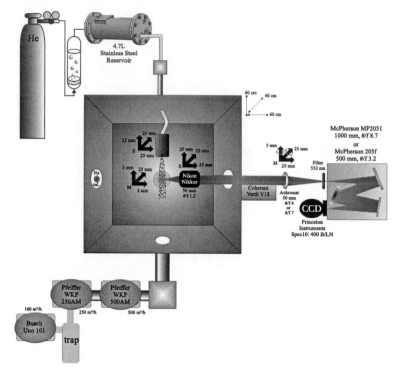

Figure 2.1: Illustrative representation of the structure of the *curry*-jet in a stylized top view, taken from Ref. 39.

$60 \times 60 \times 40 \, \mathrm{cm}^3$. The supply line between saturator and reservoir is controlled by pulsed solenoid valves (Asco Joucomatic, G262.544 V) with adjustable opening and closing times between 0.1 and 10 s. Multi-component mixtures of various concentration ratios can be created by controlling the three independent supply lines connected with the reservoir. The expansion is controlled via a solenoid valve (Parker Lucifer, type 221J3301E-299560-483816C2) and synchronized with the shutter of the CCD camera. The stagnation pressure p_s in the reservoir and the background pressure p_b in the chamber can be varied to influence the expansion mechanism [54]. By increasing the background pressure, the shock waves can also be shifted to positions close to the laser focus, thus giving rise to spectra of warmer clusters. The chamber is evacuated by a $250\,\mathrm{m}^3/\mathrm{h}$ Roots pump (Pfeiffer Vacuum, type WKP 250 AM) backed by a $100\,\mathrm{m}^3/\mathrm{h}$ rotary vane pump (Dr.-Ing. K. Busch GmbH, type UNO 101 S). In most cases, a $500\,\mathrm{m}^3/\mathrm{h}$ Roots pump (Pfeiffer Vacuum, type WKP 550 AM) is added to further extend the zone of silence in the expansion chamber. A detailed description of the chamber and the pump system can be found in Ref. 39.

The nozzle is interchangeable. Three homebuilt slit nozzles with different geometries $(8.0 \times 0.05\,\mathrm{mm}^2,\ 4.0 \times 0.15\,\mathrm{mm}^2$ and $2.5 \times 0.20\,\mathrm{mm}^2)$ are used in this work. The nozzle can be positioned in x- and y-direction by DC motors (Newport

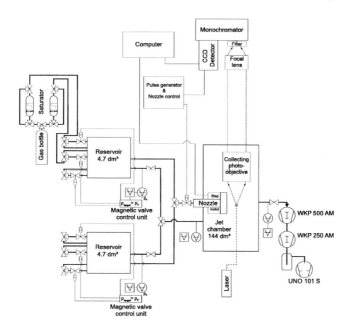

Figure 2.2: Schematic representation of the function principle of the *curry*-jet, taken from Ref. 39 and based on Fig. 3.2 from Ref .33.

GmbH, high resolution DC motor, model CMA-25CC, smallest movement interval $< 0.5\,\mu$m, displacement $25\,$mm, speed $50\text{-}400\,\mu$m/s). In z-direction (propagation direction of the expansion) the distance between the nozzle exit from laser beam is controlled through a high-precision motor drive (Newport GmbH, type LTA-HS, smallest increment $0.1\,\mu$m, resolution $0.035\,\mu$m).

A frequency doubled continuous Nd:YVO$_4$ laser ($532\,$nm, Coherent Verdi V18, $P = 18\,$W) serving as light source is focused by an adjustable lens (LINOS Photonics GmbH, type G312-300-322, plano-convex lens, $\varnothing = 22.4\,$mm, $f = 50\,$mm, ARB2 coated, $\lambda = 450\text{-}700\,$nm) on the jet expansion at a distance between $0.4\,$mm and $3\,$mm from the nozzle exit. An introduction to the laser as well as the determination of the zero position of the nozzle can be found in Ref. 39. The 90° scattered light in the chamber is collected and collimated by a short focal length lens (Nikon GmbH, Nikkor $50\,$mm $f/1.2$), which can be positioned in x-, y- and z-directions by external motors (Newport GmbH, high-resolution DC motor, model CMA-25CC, smallest movement interval $< 0.5\,\mu$m, displacement $25\,$mm, speed $50\text{-}400\,\mu$m/s) to adjust the beam path. The collimated light leaves the chamber through a quartz window (LINOS Photonics GmbH, plate with high level flatness and parallelism, type G390-117-322, ARB2 coated, $\lambda = 450\text{-}700\,$nm) and is focused on the entrance slit of the monochromator by an achromatic lens (MP2051: Edmund Optics GmbH, type L45-354, $\varnothing = 50\,$mm, $f = 350\,$mm $f/7$, MgF$_2$ coated; 205f: Newport GmbH, type PAC087AR.14, $\varnothing = 50.8\,$mm, $f = 200\,$mm $f/3.9$, AR.14 coating $430\text{-}700\,$nm) fitting to the monochromator aperture ratio. The Rayleigh scattering and scatter artifacts are filterd by an edge filter (L.O.T.-Oriel GmbH & Co. KG, REFUS532-25, USLR Raman-edge filter $\varnothing = 25\,$mm, laser line $532.0\,$nm, OD 6.0, transmission range $535.4\text{-}1200\,$nm, $T > 90\%$ on average) in front of the monochromator. A careful alignment of the filter is important for measurements near the Rayleigh line.

Two different monochromators are available for the analysis of the scattered light: the McPherson Inc. Model 2051 (MP2051, built in 1978, $f/8.6$, $f = 1000\,$mm, grating $1200\,$gr/mm $110 \times 110\,$mm^2) and a McPherson Inc. Model 205F ($f/3.2$, $f = 500\,$mm, a grating with $600\,$gr/mm (blaze $500\,$nm), $f/3.6$ and a holographic master grating, 1800 gr/mm, $f/3.4$, both Al-coated). The 205f reaches a better light intensity but a worse resolution. All the spectra reported in this work were carried out with MP2051.

The dispersed light is detected by a back-illuminated liquid N_2 cooled CCD camera (PI Acton, Spec-10: 400B/LN, 1340×400 Pixel), followed by blockwise averaging and automated removal of cosmic radiation events. Calibration with Ne lines yields the spectra displayed in the results section. The spectra represent averages over several 30-600 s jet measurements. A measurement in vacuum under otherwise identical conditions is often used as a reference background spectrum. Cosmic ray signals are removed iteratively by comparing these measurement blocks using a program named SpecTool written by P. Zielke [33]. For the measurements in the low wavenumber region (0-700 cm^{-1}, i.e., the intermolecular fundamental region of the carboxylic acid dimers), normally no background correction is carried out on the CCD readout, because the jet expansion displaces any residual air impurities in the chamber. In some cases, neat He expansions were used as a reference without inelastic scattering for checking purposes. The obtained spectra have to be calibrated to convert the pixel number into wavelength. A spectrum of a Ne-emission lamp (L.O.T.-Oriel GmbH & Co. KG, line source for spectral calibration, neon, type LSP032) is used for that purpose.

As mentioned in Chap. 1.1.2, depolarization measurements are carried out by rotating the excitation laser from its perpendicular polarization relative to the scattering plane into a parallel orientation using a $\lambda/2$ retardation plate (Edmund Optics GmbH, type E43-695, $\lambda = 532$ nm). Bands which are reduced in intensity by more than a factor of $6/7$ are due to totally symmetric vibrations, according to Eq. 1.24.

Due to the continuous improvement and test of *curry*-jet, different components of the setup (nozzle/reservoir/pumps) are used in different time periods. For the jet experiments, the 2.5×0.2 mm^2 nozzle, the 4.7 L Teflon-coated reservoir and all the three pumps were used unless indicated otherwise.

2.2 Band position error

Wavenumber differences around $1\,\mathrm{cm}^{-1}$ for the same spectral bands measured in the IR and Raman jet measurements are found, mainly in the high wavenumber region (above $3000\,\mathrm{cm}^{-1}$). This deserves a further discussion because these shifts can be found in the measurements of many different compounds, although there are differences between the expansion environments of the two experimental setups (for example, the vibrational temperature in the *filet*-jet is much lower than that in the *curry*-jet) and even with the same setup, different measurement conditions (for example, with different compounds concentration) may cause band shifts. Several IR/Raman band positions are listed as examples in Tab. 2.1. They are mainly from unpublished measurement results in our group. No exact rule is found, but the Raman band positions are always higher than those from IR measurements. One reason may be imperfection in the optical element alignment of the *curry*-jet. Another error source could come from the conversion process from the pixel number into wavelength.

Band description	IR	Raman	$\triangle\tilde{\nu}_{(\mathrm{Raman - IR})}$	Calibration set (Ne)
ν(O–H), neopentyl alcohol (M)	3676.2	3676.9	0.7	A
ν(O–H), neopentyl alcohol (M)	3667.8	3668.7	0.9	A
ν(O–H), methyl lactate (M)	3565.3	3566.0	0.7	B
ν(N–H), pyrrolidin (M)	3324.8	3325.9	1.1	B
ν(N–H), pyrrolidin (D)	3260.7	3261.3	0.6	B
β−Propiolactone[a]	1882.3	1883.8	1.5	C
ν(C=O), β-propiolactone (M)	1871.6	1872.3	0.7	C
β-Propiolactone[a] (M)	1834.0	1835.7	1.7	C
β-Butyrolactone[a] (M)	1884.1	1884.9	0.8	C
ν(C=O), β-Butyrolactone (M)	1862.0	1862.5	0.5	C
ν(C–O), methyl lactate (M)	1269.1	1269.4	0.3	D

[a] Fermi resonance band with ν(C=O).

Table 2.1: Comparison of the wavenumbers (in cm^{-1}) of several bands observed in the IR and Raman jet measurements. M means monomer and D dimer. The Raman band positions are generally the average values from a group of measurements. Different sets of spectral lines of a Ne-emission lamp were used for the calibration: A (667.82766 nm, 665.20925 nm, 659.89528 nm, 653.28824 nm, 650.65277 nm), B (659.89528 nm, 653.28824 nm, 650.65277 nm, 644.47118 nm), C (597.55343 nm, 594.48340 nm, 588.18950 nm, 585.24878 nm, 580.44496 nm), D (568.98163 nm, 566.25489 nm, 565.66588 nm, 565.25664 nm, 556.2766 nm).

2.3 Influence of the measurement conditions on the spectra

In the jet measurements of carboxylic acid dimers, the experimental conditions like the acid concentration as well as differences in the expansion environment lead to slightly different band shapes and band positions due to varying contributions by aggregates of dimers [4, 54, 55]. The dimers themselves aggregate further in the carrier gas due to dispersion forces, on the way to aerosols [55], and this contribution varies for different kinds of acids. For the case of formic acid dimer, the changes in band maxima were normally less than the spectral resolution of our setup (about 1-1.5 cm^{-1} in different wavenumber regions) [4], although larger clusters are undoubtedly present at the highest concentrations.

Fig. 2.3 shows the comparison of the Raman jet spectra of $(HCOOH)_2$ under various measurement conditions. Both in the intermolecular and the intramolecular region, the band positions are not obviously changed, with different compound concentration, nozzle distance d and stagnation pressure p_s. However, too high band intensities cause band overlap if the bands are too close to each other. Besides, measurements under different conditions afford more background information for the band assignments. Therefore, the jet spectra of the acids in this work were generally measured under at least two different conditions.

In contrast, for the case of acetic acid dimer, the obvious difference of the band shapes as well as band position shifts caused by different measurement conditions, especially for the intermolecular modes, cannot be neglected any more [54]. Another sensitive mode is the C=O stretching mode, whose wavenumber can change up to 10 cm^{-1} under extremely different expansion conditions. A possible explanation for the difference between the two acid dimers involves the small size and planarity of the formic acid dimer, which allows for vibrational motion without too much interference with neighboring molecules. Therefore, in this work the accurate dimer fundamental band positions of the acetic acid dimer were measured under the conditions avoiding aggregates of dimer (**Condition I**): The saturator temperature was set to 10°C and the resulting acetic acid concentration in the gas

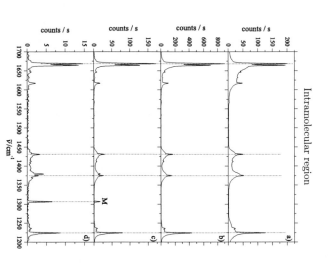

Figure 2.3: Comparison of the Raman jet spectra of $(HCOOH)_2$ under different measurement conditions. For the spectra between 100 and 700 cm^{-1}: a) $T_s = 16°C$ (3.7% HCOOH in He), $p_s = 700$ mbar, $d = 1$ mm, 4×100 s, b) $T_s = 16°C$, $p_s = 700$ mbar, $d = 0.5$ mm, 4×100 s, c) $T_s = 16°C$, $p_s = 200$ mbar, 4×200 s, d) $T_s = -10°C$ (0.7% HCOOH in He), $p_s = 700$ mbar, $d = 0.5$ mm, 4×300 s. No background correction was carried out. For the spectra between 1200 and 1700 cm^{-1}: a) $T_s = 16°C$, $p_s = 700$ mbar, $d = 1$ mm, 8×120 s, b) $T_s = 16°C$, $p_s = 700$ mbar, $d = 0.5$ mm, 8×60 s, c) $T_s = 16°C$, $p_s = 200$ mbar, $d = 0.5$ mm, 8×120 s, d) $T_s = -10°C$, $p_s = 700$ mbar, $d = 0.5$ mm, 8×300 s. M indicates monomer signals.

mixture was estimated at 0.7% (see Tab. 2.3 later on). The stagnation pressure in the reservoir was set to 200 mbar. A $2.5 \times 0.2\,\mathrm{mm}^2$ slit nozzle was used for the expansion with a short nozzle distance of 0.5 mm.

Fig. 2.4 shows a comparison of the Raman jet spectra of the OCO in-plane bending fundamental of $(CH_3COOH)_2$ under two different measurement conditions. The top one (trace a) was taken under **Condition I** whereas the other one under the condition with higher cluster abundance (**Condition II**, trace b). In this case the stagnation pressure was raised to 700 mbar and a $4 \times 0.15\,\mathrm{mm}^2$ slit nozzle was used with nozzle distance of 0.5 mm. The saturator temperature was set to 20°C and resulted in a concentration of ~1.7% acid in helium. An advantage of this condition is that intense signals allow to observe some bands which were too weak to observe under **Condition I**, although the band positions are already shifted. Under **Condition II** not only the band shapes are obviously different but also a blue shift $> 2\,\mathrm{cm}^{-1}$ of the band position could be observed relative to dimer dominant conditions. Therefore, normally the spectra measured under **Condition II** are only used for comparison but not for the fundamental band positions in this work.

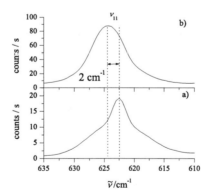

Figure 2.4: Comparison of the Raman jet spectra of $(CH_3COOH)_2$ under two extremely different measurement conditions. a) **Condition I**, $6 \times 200\,\mathrm{s}$; b) **Condition II**, $12 \times 300\,\mathrm{s}$.

2.4 Chemicals and gases

The most important compounds used in this work are presented in Tab. 2.2.

Compound	Company	Purity
He	Air Liquide	99.996%
He	Linde	99.996%
Ne	Air Liquide	99.99%
Ar	Air Liquide	99.998%
Ar	Linde	99.998%
HCOOH	Sigma-Aldrich	98%
HCOOH	Acros Organics	99%
HCOOD	Cambridge Isotope Lab.	98% D
HCOOD	Sigma-Aldrich	95% D
DCOOH	Cambridge Isotope Lab.	98% D
DCOOH	Sigma-Aldrich	95% D
DCOOD	Cambridge Isotope Lab.	98% D
DCOOD	Sigma-Aldrich	95% D
CH_3COOH	Merck	100%
CH_3COOH	Acros Organics	99.8%
CH_3COOD	Sigma-Aldrich	98% D
CD_3COOD	Sigma-Aldrich	99.5% D
HCCCOOH	Acros Organics	98%
$C(CH_3)_3COOH$	Fluka	99%
H_2O	Acros Organics	99.999%
D_2O	Euriso-Top	99.9%

Table 2.2: Compounds used in this work.

2.5 Determination of the compound concentration

The concentrations of the studied compounds are estimated according to the equation:

$$\frac{p/\text{bar}}{p_0/\text{bar}} \cdot 100 = C. \tag{2.1}$$

p is the vapor pressure at the given temperature of the saturator, p_0 the pressure of the carrier gas and C the concentration in percent.

For most measurements in this work, helium was used as the carrier gas and p_0 was normally set to 1.5 bar. The pressure dropped to $\sim$1.0 bar during the filling of the reservoir, but did not stay constant. Therefore, no true thermodynamic equilibrium between the liquid/solid phase (compound) and the gas flow could be arrived, and so p_0 was assumed to be 1 bar for all these cases. In addition, the flow speed of different carries gases and the filling height of the saturator also play a role. Therefore, only the "maximal" but no "accurate" concentration of the compound could be assumed and the following approximate equation was used:

$$p/\text{bar} \cdot 100 = C. \tag{2.2}$$

The vapor pressure can be calculated with the Clausius-Clapeyron equation but involves a relatively large error. For some compounds, corrected equations (Antoine equations) in a restricted temperature range have been published and provide more accurate vapor pressures. For the calculations of the vapor pressures in the range of measurement temperatures of this work, the following equations are used.

For formic acid and its deuterated isotopomers between 0°C and 24°C [56]:

$$\log(p/\text{bar}) = 2.00121 - \frac{515.000}{T/\text{K} - 139.408}. \tag{2.3}$$

For acetic acid and its deuterated isotopomers below 17°C [57]:

$$\ln(p/\text{Pa}) = 52.3271 - \frac{10205}{T/\text{K}} - 0.03441 \cdot T/\text{K}. \tag{2.4}$$

For acetic acid and its deuterated isotopomers over 17°C [58]:

$$\log(p/\text{bar}) = 4.68206 - \frac{1642.540}{T/\text{K} - 39.764}. \tag{2.5}$$

For H_2O and D_2O between 0°C and 30°C [59]:

$$\log(p/\text{bar}) = 5.40221 - \frac{1838.675}{T/\text{K} - 31.737}. \tag{2.6}$$

The pivalic acid was measured at 16°C and 24°C in this work. Its vapor pressures at these two temperatures are unknown. This compound has a melting point of 35.5°C and the vapor pressure at 25°C is 0.72 mbar [60]. The concentration of pivalic acid is assumed as 0.04% at 16°C and 0.07% at 24°C. The same case happened by propiolic acid. Its vapor pressure is 4.9 mbar at 25°C (production information) and so the concentration should be less than 0.5% at 16°C.

For the carboxylic acids at elevated concentrations, another complication applies. The vapor pressure involves a mixture of monomer and dimer, which depends strongly on temperature. Therefore, the estimated concentrations become particularly uncertain. Furthermore, the use of the same equation for different isotopomers introduces a certain error. However, the systematic error due to incomplete saturation is expected to dominate and the concentrations should only be viewed as relative indications of number density. That is also why some vapor pressures are calculated at temperatures out of the validity-range of the vapor pressure equation (see Tab. 2.3).

$T/^\circ$C	$p/$mbar	maximal concentration
HCOOH and its deuterated isotopomers		
-20^a	3.0	0.3%
-15^a	4.6	0.5%
-10^a	6.9	0.7%
-8^a	8.0	0.8%
-5^a	10.0	1.0%
-2^a	12.4	1.2%
0	14.1	1.4%
8	23.3	2.3%
9	24.7	2.5%
15	34.6	3.5%
16	36.5	3.7%
20	44.8	4.5%
22	49.5	5.0%
CH_3COOH and its deuterated isotopomers		
8	5.8	0.6%
10	6.9	0.7%
16	12.4	1.2%
20	15.8	1.6%
HCCCOOH		
16	< 4.9	$< 0.5\%$
$C(CH_3)_3COOH$		
16	< 0.72	$\sim 0.04\%$
24	~ 0.72	$\sim 0.07\%$
H_2O/D_2O		
22	26.4	2.6%

^{a} Temperature out of the validity-range of the vapor pressure equation.

Table 2.3: Estimated concentration list of the compounds measured in this work at all the measurement temperatures.

3 Formic acid

Two hallmarks of the hydrogen bond are its directionality and its cooperativity. The former is a key to molecular recognition [61], whereas the latter has consequences for fundamental dynamical processes like concerted hydrogen transfer [62]. Both aspects are of paramount importance in biology but they also need to be characterized at a structurally more elementary level in order to advance a quantum description of biomolecular processes.

Dimers of carboxylic acids are among the most elementary systems which exhibit accentuated hydrogen bond directionality and some degree of cooperativity. Formic acid (FA) is the simplest one in the group. Following the strategy of firstly studying building blocks and then their pair-wise interactions, the various types of monomer and dimer of HCOOH and its deuterium isotopomers are experimentally and theoretically investigated. Spontaneous Raman spectroscopy in a jet and in the gas phase (GP) at room temperature serves to analyze the vibrational dynamics of the acid monomer as well as the different kinds of dimers.

There are two conformers of the FA monomer molecule, which differ by the position of the two hydrogen atoms: *trans*-FA with a 180° HCOH angle and *cis*-FA with a 0° angle. The *cis*-FA is 16.7 kJ/mol energetically less stable [65]. The *trans*-FA is the predominant naturally occurring form in the GP: the Boltzmann population ratio between the *trans* and *cis* monomer at 298 K is about 846:1 and it increases very rapidly when the temperature is even lower. The torsional barrier between them is very high (ca. 50 kJ/mol), and difficult to overcome by thermal annealing. Besides, thermal excitation of *trans*-FA results in a direct dissociation of the molecule. Therefore, the *cis*-FA monomer has only been observed in the Ar matrix by vibrational excitation of the *trans* monomer [64]. In our experiment, both in the jet and in the GP, the *cis*-FA will not be regarded as a relevant candidate for the band assignment.

A variety of hydrogen-bonded dimers (FAD) can be constructed from the two monomer rotamers, several of which are shown in Fig. 3.1. Among them, the cyclic *trans-trans* dimer (FFa) with a C_{2h} point group symmetry is the most stable one and therefore attracted the most attention, experimentally as well as theoretically. The acyclic *trans-trans* dimer (FFb) is the second most stable, about 5.6 kcal/mol energetically higher than FFa according to a BSSE corrected quantum mechanics calculation at the MP2/6-311++G(3df,2p) level [63]. The IR absorption spectrum of FFb in an Ar matrix at 8.5 K was reported in 2006 [66]. The fraction of FFb is expected to increase with increasing temperature and pressure. Very recently, a

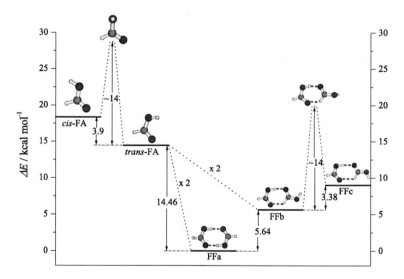

Figure 3.1: Energy diagram of the different conformers of FA and FAD. The energy differences between the dimers are from Ref. 63, which are calculated at the MP2/6-311++G(3df,2p) level with BSSE correction. The energy difference between the two monomer conformers and the torsion barrier values are from Ref. 64.

weak band at $864 \pm 2.1\,\mathrm{cm}^{-1}$ observed in the GP Raman spectrum was assigned as the $\gamma(\text{O–H})$ vibration of FFb [63].

The *cis-trans* dimer (FFc) shown in Fig. 3.1 can be produced by light-induced rotation of the free OH bond of FFb. Like in the monomer, the torsional barrier is ca. $50\,\mathrm{kJ/mol}$. Several bands were detected using Ar matrix isolation [66]. FFc and other dimers with similar interaction energies will not be considered for the assignments of the bands that are observed with our methods.

With the help of anharmonic calculations, attempts have been made to search for the metastable FFb conformation. Due to the much lower vibrational temperature, it is much more difficult to observe the FFb dimer in the jet spectra. Therefore, the *trans*-FA monomer and possible FFb bands are analyzed at first in the GP spectra, with comparison to the jet spectra. After that, the fundamental as well as overtone/combination bands of FFa in the jet spectra will be discussed in three wavenumber regions: the low-frequency intermolecular vibration region ($\leq 750\,\mathrm{cm}^{-1}$), the finger-print intramolecular vibration region (750-1800 cm^{-1}) and the O–H/D as well as C–H/D vibration region ($\geq 1800\,\mathrm{cm}^{-1}$). Anharmonicity aspects will be discussed based on accurate fundamental band positions and quantum mechanics calculations. A rigorous experimental reference frame for existing and future high level quantum chemical and dynamical treatments of this important prototype system is thereby provided.

3.1 Formic acid monomer

Unlike the complex and highly congested jet spectra due to the complicated intermolecular interactions, the GP spectra of the formic acid are much simpler and will be discussed in detail at first.

3.1.1 HCOOH monomer

Comparisons of the Raman jet and GP spectra of HCOOH (with monomer-dimer equilibria) in the wavenumber range of 100-3600 cm^{-1} are shown in Fig. 3.2, separated in three regions for a better view. The spectra are combinations of measurements in different wavenumber regions but under equivalent conditions, because the measurement window for the $1200 \, \text{G/mm}$ grating varies from $550 \, \text{cm}^{-1}$ to $700 \, \text{cm}^{-1}$ in different wavenumber regions, with the McPherson f/8.6 1000 mm monochromator. The labels in the GP spectra correspond to monomer vibrations and the labels in the jet spectra to those of the dimer.

As already mentioned, the bands in the GP spectra are mainly from the *trans*-FA monomer whereas the bands of FFa are obviously dominant in the jet spectra. The monomer bands in the jet spectra are generally very weak at high compound concentration. The fundamental bands have been numbered in the spectra following the Herzberg nomenclature [8] (from ν_1 to ν_9 for the fundamentals of *trans*-FA in the GP spectra and from ν_1 to ν_{12} for the Raman active fundamentals of FFa in the jet spectra), which was used in the most references. The fundamental modes of *trans*-FA from ν_1 to ν_7 with A′ symmetry are ordered from higher to lower wavenumber, whereas the remaining two (ν_8 and ν_9) have A″ symmetry. FFa has 24 fundamental modes which may be classified according to their C_{2h} point group symmetry behavior and which correspond to 13 different kinds of vibrational motion. While the intermolecular twisting mode ν_{16} (in Herzberg notation, adopting, however, the representation sequence A_g, B_g, A_u, B_u [9]) is only IR active and the intermolecular stretching mode ν_8 is only Raman active, the other 11 vibrations occur in Davydov pairs of IR active and Raman active modes, which are split in a mode- and isotope-dependent way [67]. Detailed descriptions of the modes can be found in the following tables.

The fundamental wavenumbers of the *trans*-HCOOH have been reported in early low resolution IR gas phase work [18,71,75] and were re-investigated with Raman GP spectroscopy [9,67] and with high-resolution IR rovibrational analysis recently [73]. IR [64,68] as well as Raman spectra [69] in Ar matrix were also reported. These previous experimental results are summarized in Tab. 3.1, compared with the measured band maxima (from Fig. 3.2) and assignments of this work. For the band description, ν means bond stretching, δ valence angle bend-

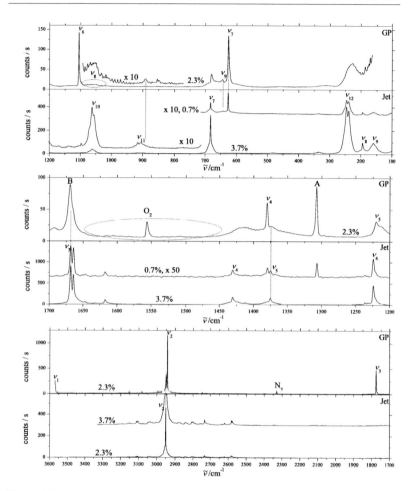

Figure 3.2: Raman spectra of HCOOH between 100 and 3600 cm^{-1}. The jet spectra were generally measured under two different conditions: a), $T_s = 16°C$ (3.7% HCOOH in He), $p_s = 700$ mbar, $d = 0.5$ mm, 8×100-200 s (4×100 s for the spectrum between 100 and 700 cm^{-1}); b), $T_s = 8°C$ (2.3% HCOOH in He), $p_s = 500$ mbar, $d = 1$ mm, 8×100-300 s. A jet spectrum with relatively low HCOOH concentration in the low wavenumber region is shown for comparison: $T_s = -10°C$ (0.7% HCOOH in He), $p_s = 700$ mbar, $d = 0.5$ mm, 4×300 s. The GP spectra correspond to 2.3% HCOOH in He ($T_s = 8°C$), the background pressure in the chamber $p_b = 30$ mbar, 4×30 s.

		GP			Ar matrix	
Mode	Description	This work	IR	Raman[a]	IR[b]	Raman[c]
ν_1 (A')	ν(O–H)	3570	3570[d,e]	3568.9	3550.5/3548.2	3553.5
ν_2 (A')	ν(C–H)	2943	2943[d], 2944[e]	2942.0	2956.1/2953.1	2958.5
ν_3 (A')	ν(C=O)	1777	1770[d], 1776[e]	1776.6	1768.9/1767.2	1769.0
ν_4 (A')	δ(C–H)	1380	1387[d], 1380[f], 1368.99[g]	1380.6	1384.4/1381.0	1386.0
ν_5 (A')	δ(O–H)	1220	1229[d], 1223[e], 1248.67[g]	. . .	1215.8/1214.8	1218.0
ν_6 (A')	ν(C–O)	1105	1105[d,e,h], 1109.20[g]	1103.8	1106.9/1106.8	1107.2
ν_7 (A')	δ(OCO)	626	636[d], 626.2[i]	624.9	629.3/628.0	626.0
ν_8 (A'')	γ(C–H)	1045-1067	1033[d,e,h], 1033.47[g]	. . .	1038.5/1037.4	1040.0
ν_9 (A'')	γ(O–H)	643	636[d], 640.8[i]	642	635.4	632.5

[a] Bertie and Michaelian, Ref. 9
[b] Maçôas et al., Ref. 68
[c] Olbert-Majkut et al., Ref. 69
[d] Millikan and Pitzer, Ref. 70
[e] Hisatsune and Heicklen, Ref. 71
[f] Redington and Lin, Ref. 72
[g] Baskakov et al., deperturbed band centers, Ref. 73
[h] Williams, Ref. 10
[i] Deroche et al., Ref. 74

Table 3.1: Comparison of the band positions (in cm^{-1}) of the fundamental modes of *trans*-HCOOH.

ing and γ out-of-plane bending. These symbols will be used further on for the following tables.

When comparing the present work data to the experimental values from Bertie *et al.* (Raman GP spectrum of HCOOH at 21°C) [9], the bands positions are in very good agreement and all the wavenumber differences are within 1.2 cm^{-1}.

The two modes not observed in their work deserve detailed discussion: ν_5 and ν_8. The broad band at 1220 cm^{-1} in the GP spectrum is assigned as ν_5 in this work, which fits well to the band positions of the former studies [68–70,74]. The band at 1308 cm^{-1} (band A) is very strong in our GP spectrum and relatively weak in the jet spectra, and almost disappears in the jet spectrum with a relative high acid concentration of 3% in He (see Fig. 3.2). This is a typical behavior of the relative band intensities of a monomer band from GP to jet spectra (with increasing compound concentrations). This band was observed at 1307.1 cm^{-1} in Ref. 9 and assigned as the overtone of ν_9 of the monomer. This assignment was surpported by the high resolution study on the $2\nu_9 \leftarrow \nu_9$ hot band [73]. Its anomalously high intensity and the relatively large positive anharmonicity are explained by a strong Fermi resonance between $2\nu_9$ and ν_5 [73].

The C–H out-of-plane bending mode ν_8 was not observed in Ref. 9 due to its very weak Raman activity. Even in our spectrum only a broad band is observed but no dominant single peak can be distinguished.

The band at $\sim$1670 cm^{-1} in the GP spectrum (marked with Label B, see Fig. 3.2) is very strong and broad. Contributions from the HCOOH-H$_2$O-complex cannot be ruled out. A further measurement series of the admixture of HCOOH and H$_2$O may reveal the true nature of this band.

A comparison of the experimental and theoretical band positions of the fundamental modes of the *trans*-HCOOH is shown in Tab. 3.2. It is seen that the band positions from the harmonic calculations using different methods at different levels are generally overestimated. The anharmonic calculation at the MP2/6-311+G(2d,p) level has a best agreement with most experimental values. ν_5 is predicted at 1253.6 cm^{-1} at this level and this supports the assignment mentioned above, although all the harmonic calculations predict a band position around 1307 cm^{-1}.

Significant wavenumber differences exist between the experimental value and the theoretical band positions of the O–H out-of-plane bending mode ν_9: all the harmonic calculations predict a band position around 680 cm^{-1} whereas all the anharmonic calculations give wavenumbers below 630 cm^{-1}. Both do not fit well to the observed band position of 643 cm^{-1} in this work and the similar wavenumbers around 640 cm^{-1} from the former studies using different methods: 642 cm^{-1} in the Raman GP spectra [9], 636 cm^{-1} in the low resolution IR GP spectra [70], 640.7 cm^{-1} with the high resolution FTIR experiment [74], 635.4 cm^{-1} in the IR absorption spectra in an Ar matrix [68] and 632.5 cm^{-1} in the recent Raman spectra in Ar matrix [69]. It is furthermore interesting that there is just a small peak at 686 cm^{-1} in the spectra, on the blue-side shoulder of ν_7 of FFa at 682 cm^{-1}, which deserves a further discussion.

Further experiments with various jet/GP conditions are done to verify these two bands. Fig. 3.3 shows a continuous spectral transition between jet-cooled conditions (upper traces) and conditions in which the dimers are heated during their deceleration in the receding Mach disk zone, by increasing p_b in the jet expansion for a given d and p_s. Another group of the Raman jet spectra under the same measurement conditions except with 3.7% HCOOH in He is shown in Fig. 3.4.

The band assigned as ν_{11} by Bertie *et al.* at $642\,\mathrm{cm}^{-1}$ was observed at $643\,\mathrm{cm}^{-1}$ in all the spectra. The band at $686\,\mathrm{cm}^{-1}$ can be observed also in the jet spectra but only distinguished in the spectra at much lower FA concentration in He (0.7%) because this band is so close to the intense ν_7 mode of FFa at $682\,\mathrm{cm}^{-1}$.

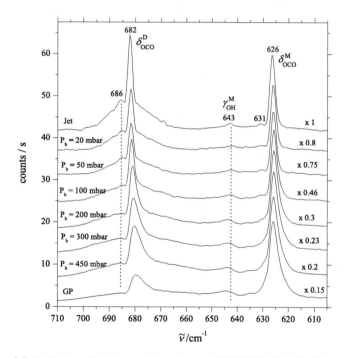

Figure 3.3: Evolution of the Raman jet spectrum of 0.7% HCOOH in He ($T_s = -10°C$) with increasing background pressure, scaled to similar scattering intensity of the monomer band at $626\,\mathrm{cm}^{-1}$, $p_s = 700\,\mathrm{mbar}$, $d = 0.5\,\mathrm{mm}$, $4 \times 50\text{-}200\,\mathrm{s}$. A 300 K gas phase spectrum ($T_s = -10°C$, $p_b = 30\,\mathrm{mbar}$, $4 \times 50\,\mathrm{s}$) is shown for comparison at the bottom.

Mode	Experiment	B3LYP			MP2			
		6-31+G*	6-311+G*	6-311++G(2d,2p)	6-31+G*	6-311+G*	6-311+G(2d,p)	6-311+G(3d,p)
A'								
ν_1	3570	3664.7	3701.2	3741.1 (3537.9)	3658.7	3744.5	3746.0 (3559.6)	3745.8 (3559.9)
ν_2	2943	3103.6	3071.0	3057.5 (2894.2)	3176.6	3139.9	3115.0 (2949.8)	3118.9 (2953.3)
ν_3	1777	1822.9	1819.3	1809.0 (1775.5)	1797.8	1807.5	1784.1 (1751.6)	1788.7 (1755.6)
ν_4	1380	1408.6	1410.9	1403.1 (1387.0)	1428.0	1428.9	1419.9 (1397.0)	1413.7 (1394.5)
ν_5	1220	1299.6	1297.6	1305.8 (1255.1)	1307.0	1312.9	1309.4 (1253.6)	1298.5 (1195.6)
ν_6	1105	1136.8	1132.4	1120.7 (1088.9)	1139.8	1152.3	1120.6 (1087.8)	1125.0 (1091.9)
ν_7	626	622.5	629.5	628.8 (623.0)	621.5	634.2	632.1 (624.7)	625.3 (619.7)
A''								
ν_8	1045-1067	1046.6	1046.4	1050.5 (1031.0)	1055.9	1062.2	1053.0 (1032.7)	1052.3 (1031.4)
ν_9	643	690.5	694.8	675.6 (613.6)	688.4	693.6	673.7 (611.5)	676.9 (629.9)

Table 3.2: Comparison of the experimental (this work) and theoretical band positions (in cm^{-1}) of the fundamental modes of the *trans*-HCOOH. The anharmonic calculation results are listed in parentheses.

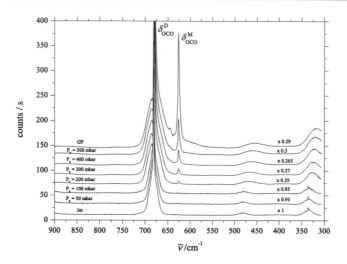

Figure 3.4: Evolution of the Raman jet spectrum of 3.7% HCOOH in He ($T_s = 16°C$) with increasing background pressure, scaled to similar scattering intensity of the FFa band at 682 cm^{-1}, $p_s = 700$ mbar, $d = 0.5$ mm, 4×30-100 s. A 300 K gas phase spectrum ($T_s = 16°C$, $p_b = 30$ mbar, 4×20 s) is shown for comparison on the top.

The band at 642 cm^{-1} cannot correspond to FFa or larger clusters, because it is not observed in the jet spectra of high concentration (see Fig. 3.2), which is the preferential condition for the energetically most stable FFa. Besides, the band is both IR and Raman active. No symmetry selection rule like in the C$_{2h}$ point group of FFa applies. It can be observed in an Ar matrix at 8.5 K [66], so it should not be a hot band. A mixed complex with water is also unlikely. With the comparison of the spectra in the two figures it is found that this band keeps a specific ratio with the monomer band at 626 cm^{-1}. This relationship is shown in Fig. 3.5 and Fig. 3.6 in more detail: the relative intensity of the FFa band at 680 cm^{-1} in the lower acid concentration GP spectrum ($T_s = -10°C$, see Fig. 3.6) decreases a lot compared to that in the higher acid concentration GP spectrum ($T_s = 16°C$, see Fig. 3.6), but the intensity of the band at 642 cm^{-1} stays constant

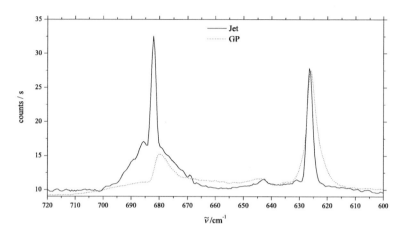

Figure 3.5: Comparison of the Raman jet and GP spectra of 0.7% HCOOH in He ($T_s = -10°C$), scaled to similar scattering intensity of the monomer band at $626\,cm^{-1}$. Both spectra are taken from Fig. 3.3.

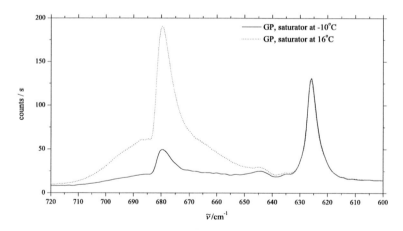

Figure 3.6: Comparison of the Raman GP spectra with different concentrations of HCOOH in He, scaled to similar scattering intensity of the monomer band at $626\,cm^{-1}$. The spectra are taken from Figs. 3.3 and Figs. 3.4.

within the spectrum base line accuracy, when the monomer band at $626\,\mathrm{cm}^{-1}$ is scaled to similar scattering intensity as reference. Due to this band intensity relationship the band at $642\,\mathrm{cm}^{-1}$ cannot belong to another dimer FFb which is energetically less stable than FFa, although all the quantum chemical calculations, both harmonic and anharmonic, indicate that the OCO in-plane bending of FFb has a wavenumber around $645\,\mathrm{cm}^{-1}$ (see Tab. 3.3 later on). After all, it can only come from the *trans*-HCOOH monomer. The weak feature at $686\,\mathrm{cm}^{-1}$ is believed to be due to rotational structure of the dimer band at $682\,\mathrm{cm}^{-1}$.

In 2009 R. Balabin reported a weak band at $864 \pm 2.1\,\mathrm{cm}^{-1}$ in the Raman GP spectrum, which was assigned as the $\gamma(\mathrm{O\text{-}H})$ vibration of FFb [63], although it has the weakest Raman cross section among the 24 vibrational modes of FFb according to his calculations at the extensive MP2/6-311++G(3df,2p) level [63]. All the other stronger bands were not observed because they are masked by the monomer bands. A weak band is also observed in our GP spectrum, with a doublet structure at $849/854\,\mathrm{cm}^{-1}$, but not in the jet spectrum, which means that it does not belong to FFa. If this band really belonged to FFb, it would be hard to imagine the OCO in-plane bending mode with a 34 times bigger Raman cross section [63] could not be observable, with the spectral resolution of our setup. It is unlikely that this band comes from the monomer itself (fundamental, overtone/combination/difference band), but the possibility of a mixed complex of water and HCOOH could not be easily ruled out. Another band around $770\,\mathrm{cm}^{-1}$ is also visible in the spectrum in Ref. 63, much stronger than the band at $864\,\mathrm{cm}^{-1}$, but without any discussion. This band is not observed in the Raman GP spectrum of this work und therefore may also not be from formic acid itself. Analysis of the jet spectra of formic acid in the $\leq 750\,\mathrm{cm}^{-1}$ region under extremely high signal intensity conditions (see Fig. 3.15 in the next section) shows that all spectral features even down to the noise level can be basically attributed to symmetric dimers and monomers of formic acid, supporting the expectation that isomeric forms of the dimer are not present in significant amounts. Overall, there is no hard evidence for the existence of FFb in the GP spectra in this work and in the following text, FFa as the only dimer-form of FA in our spectra will be discussed exclusively.

Furthermore, there are some overtone/combination bands of *trans* monomer in the GP spectra. For example, the three intense narrow bands around ν_2 (see Fig. 3.7) can be assigned as follows: the band at $3153\,\mathrm{cm}^{-1}$ as $\nu_3 + \nu_4$ $(1777 + 1380 = 3157\,\mathrm{cm}^{-1})$, the band at $3083\,\mathrm{cm}^{-1}$ as $\nu_3 + \nu_5$ $(1777 + 1308 = 3085\,\mathrm{cm}^{-1})$ and the band at $2747\,\mathrm{cm}^{-1}$ as $2\nu_4$ $(2 \times 1380 = 2760\,\mathrm{cm}^{-1})$. All of them have small anharmonicities within $13\,\mathrm{cm}^{-1}$.

Several FFb fundamental bands have been assigned in the Ar matrix experiment [66, 69]. They are listed in Tab. 3.3, compared with calculations. An alternative assignment in terms of overtone/combination bands of *trans* monomer cannot be ruled out in all cases, if the calculations cannot make accurate theoretical predictions. Therefore, the assignment in Tab. 3.3 is very tentative. For the IR matrix experiment, the band observed at $3540.1\,\mathrm{cm}^{-1}$ is possible to be assigned as the first overtone of ν_3 $(2 \times 1777 = 3554\,\mathrm{cm}^{-1})$ and the band at $3142.8\,\mathrm{cm}^{-1}$ may

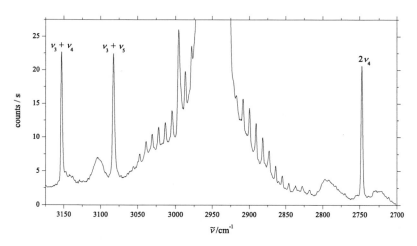

Figure 3.7: Several combination bands in the Raman GP spectra of 2.3% HCOOH in He (8°C), $p_\mathrm{b} = 30\,\mathrm{mbar}$, $4 \times 30\,\mathrm{s}$.

	Ar matrix		B3LYP		MP2			
Mode	Raman[a]	IR[b]	6-31+G*	6-311++G(2d,2p)	6-31+G*	6-311+G*	6-311+G(2d,p)	6-311+G(3d,p)
ν(O-H)	...	3540.1	3654.9	3733.2 (3533.2)	3649.3	3736.8	3738.1 (3551.0)	3738.6 (3551.8)
ν(O-H)	...	3142.8	3357.4	3374.9 (3144.2)	3425.2	3523.8	3420.3 (3231.7)	3421.9 (3233.9)
ν(C-H)	2870.0?	...	3089.9	3045.2 (2945.7)	3164.6	3129.8	3104.0 (2949.9)	3107.3 (2950.0)
ν(C-H)	...	...	3154.7	3108.3 (3045.7)	3220.4	3187.6	3163.1 (3014.7)	3166.6 (3017.6)
ν(C=O)	...	1748.2	1799.6	1783.9 (1749.5)	1786.2	1795.4	1735.5 (1701.4)	1774.5 (1739.8)
ν(C=O)	1718.0	...	1761.4	1743.1 (1707.0)	1757.1	1766.2	1766.2 (1735.5)	1740.6 (1707.1)
δ(C-H)	...	...	1436.2	1439.6 (1390.2)	1451.6	1449.3	1447.0 (1399.5)	1446.0 (1396.1)
δ(O-H)	...	...	1401.0	1401.2 (1366.7)	1421.8	1424.2	1419.1 (1401.0)	1407.6 (1411.6)
δ(C-H)	...	...	1381.1	1383.9 (1341.0)	1388.7	1386.4	1392.0 (1347.0)	1383.3 (1344.3)
δ(C-H)	...	...	1324.1	1329.3 (1283.5)	1332.1	1334.7	1333.0 (1223.6)	1322.5 (1239.7)
δ(O-H)	...	...	1213.9	1201.5 (1166.1)	1218.4	1218.3	1200.3 (1162.5)	1205.4 (1168.5)
ν(C-O)	1183.0	1180.4	1164.6	1153.5 (1119.8)	1167.9	1180.8	1152.3 (1119.1)	1154.3 (1122.1)
ν(C-O)	1136.0	1131.8	1153.5	1091.8 (1060.5)	1092.2	1095.4	1091.5 (1063.1)	1092.0 (1063.3)
ν(C-H)	...	...	1084.8	1066.0 (1048.1)	1074.1	1076.0	1069.6 (1047.7)	1074.2 (1042.6)
γ(C-H)	...	...	1062.3	917.3 (855.9)	919.9	841.9	902.0 (843.7)	927.2 (824.5)
γ(O-H)	...	867.5	918.9	697.7 (625.9)	708.9	709.9	694.3 (645.1)	696.6 (671.3)
γ(O-H)	...	...	711.8	681.3 (670.9)	697.5	694.3	680.0 (667.8)	676.6 (665.7)
δ(OCO)	669.0	658.1	697.7	647.8 (641.0)	667.5	675.1	651.8 (643.9)	646.1 (639.9)
δ(OCO)	...	...	673.7	642.7 (641.0)	643.0	652.7	651.8 (643.9)	646.1 (639.9)
γ(O-H...O)	213.0?	...	202.1	201.3 (191.9)	195.3	181.1	193.1 (178.3)	200.7 (178.9)
ν(O-H...O)	...	...	194.4	182.2 (168.7)	192.1	168.5	176.0 (162.3)	184.7 (169.8)
δ(O-H...O)	...	...	150.7	150.2 (142.6)	146.4	140.3	142.7 (135.2)	144.9 (140.2)
γ(O-H...O)	...	...	111.5	109.8 (90.6)	106.6	100.2	109.7 (98.5)	113.1 (106.4)
δ(O-H...O)	...	...	99.8	103.2 (86.4)	101.9	92.3	102.9 (91.2)	104.8 (97.1)
internol. twist	...	...	64.5	62.4 (74.6)	59.5	53.8	58.0 (54.3)	59.8 (60.7)

[a] Olbert-Majkut et al., Ref. 69.
[b] Marushkevich et al., Ref. 66.

Table 3.3.: Experimental and theoretical band positions (in cm^{-1}) and assignments of the fundamental modes of the FFb. The anharmonic calculation results are listed in parentheses.

be the same band in our GP spectra at $3153\,\mathrm{cm}^{-1}$ ($\nu_3 + \nu_4$). For the Raman matrix experiment, the band at $2870\,\mathrm{cm}^{-1}$ is possibly $\nu_3 + \nu_6$ ($1777 + 1105 = 2882\,\mathrm{cm}^{-1}$) and the band at $1718.0\,\mathrm{cm}^{-1}$ may come from ($\nu_6 + \nu_7$) ($1105 + 626 = 1731\,\mathrm{cm}^{-1}$). The assignment of bands below $1213\,\mathrm{cm}^{-1}$ is relatively certain because the lowest fundamental mode of the *trans*-HCOOH ν_7 has a wavenumber of $626\,\mathrm{cm}^{-1}$.

3.1.2 Monomers of the three deuterated formic acid isotopomers

Now the deuterated isotopomers will be discussed. In this work the Raman spectra of all the three partially and fully deuterated isotopomers were analyzed. Comparisons of the Raman GP and Jet spectra of the isotopomers are shown in Fig. 3.9 (HCOOD, $100\text{-}3200\,\mathrm{cm}^{-1}$), Fig. 3.8 (DCOOH, $100\text{-}3350\,\mathrm{cm}^{-1}$) and Fig. 3.10 (DCOOD, $100\text{-}2700\,\mathrm{cm}^{-1}$), respectively. For all the isotopomers, the spectra are separated in three wavenumber regions, and the fundamental monomer/dimer modes are numbered. The labels in the GP spectra correspond to monomer vibrations and the labels in the jet spectra to those of the dimer. All the jet spectra below $750\,\mathrm{cm}^{-1}$ were measured with the $67\,\mathrm{L}$ reservoir and the $8.0 \times 0.05\,\mathrm{mm}^2$ nozzle. Only the $250\,\mathrm{m}^3/\mathrm{h}$ Roots pump and the $100\,\mathrm{m}^3/\mathrm{h}$ rotary vane pump were used to evacuate the chamber.

It must be mentioned that in the GP spectra of HCOOD, some fundamental bands of HCOOH could be observed. The same case happened in the spectra of DCOOD for DCOOH. Such mixed bands arise mainly from partial isotope exchange from the OD-compounds with the H_2O at the container walls and therefore, evidence of isotope exchange is not seen in the HCOOH and DCOOH spectra. The band maxima and assignments are listed in Tab. 3.4, compared with some previous experimental work and the harmonic quantum chemical calculation at the simplest B3LYP/6-31+G* level, which shows already a good agreement with the experimental results of HCOOH. Most of the fundamentals of the deuterated *trans*-FA have been investigated mainly in the gas phase, using IR [18,71,75] and

Raman [9,67] spectroscopy. For DCOOH, the IR spectra in Ar matrix have been reported [68].

Most of the band positions of the four isotopomers fit very well to the experimental results of the Raman GP spectra from Bertie *et al.* [9,67], with wavenumber shifts below $3\,\mathrm{cm}^{-1}$. The wavenumber differences between our results and the IR GP as well as the Ar matrix measurement are also moderate.

The assignment of the ν_5 fundamental of HCOOH has been discussed. Similar to HCOOH, there is a relatively weak band at $1206\,\mathrm{cm}^{-1}$ and a much stronger band at $1300\,\mathrm{cm}^{-1}$ (marked with A) in the Raman GP spectrum of DCOOH (see Fig. 3.8). The latter one was observed at $1297\,\mathrm{cm}^{-1}$ in the Raman GP spectrum and assigned as the ν_5 fundamental mode of DCOOH in Ref. 67. The former one was observed at about $1200\,\mathrm{cm}^{-1}$ and assigned as ν_5 in several former IR GP studies [68,70]. Same as in the case of HCOOH, the $1300\,\mathrm{cm}^{-1}$ band is assigned at the ν_9 overtone band of the monomer and the band at $1206\,\mathrm{cm}^{-1}$ is assigned as the ν_5 fundamental of the DCOOH monomer. The high band intensity is profited from the strong Fermi resonance between $2\nu_9$ and ν_5 too, same as HCOOH.

It is not surprising that the harmonic calculation result at the B3LYP/6-31+G* level ($1295\,\mathrm{cm}^{-1}$) does not fit well the experimental band position ($1206\,\mathrm{cm}^{-1}$) but the small isotope shifts of the ν_5 band of DCOOH relative to HCOOH calculated using different methods and basis sets (all within $10\,\mathrm{cm}^{-1}$) are in good agreement with the experimental result ($14\,\mathrm{cm}^{-1}$).

For the two O-deuterated isotopomers, a good agreement is found between our experimental values and those from Bertie *et al.* [9,67]. Both bands marked with A in the spectra of HCOOH and DCOOH were also observed in the GP spectra of their related O-deuterated variants due to the 95% chemical purity of the deuterated compounds and the O–H/D exchange during the measurements. They are also marked with A in the GP spectra of HCOOD and DCOOD (see Figs. 3.9 and 3.10).

A doublet structure is observed for the C-H stretching vibration mode of HCOOD at $2943/2939\,\mathrm{cm}^{-1}$, corresponding to the experimental values of $2941.8/2938.2\,\mathrm{cm}^{-1}$ in the Raman GP spectrum in Ref. 67, in which the bands

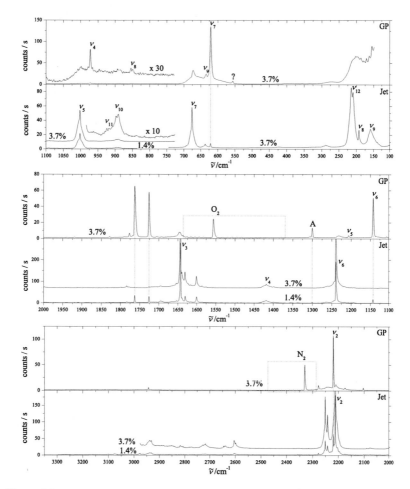

Figure 3.8: Raman spectra of DCOOH between 100 and 3350 cm^{-1}. The jet spectra were generally measured under two different conditions: a), $T_s = 16°C$ (3.7% DCOOH in He), $p_s = 700$ mbar, $d = 0.4$ mm, 8×200 s (for the spectrum between 100 and 750 cm^{-1}, $p_s = 500$ mbar, 12×240 s); b), $T_s = 0°C$ (1.4% DCOOH in He), $p_s = 700$ mbar, $d = 0.4$ mm, $8 \times 150\text{-}200$ s. The GP spectra were taken under the following conditions: $T_s = 16°C$ (3.7% DCOOH in He), $p_b = 30$ mbar, 4×60 s (6×50 s for the spectrum between 100 and 750 cm^{-1}).

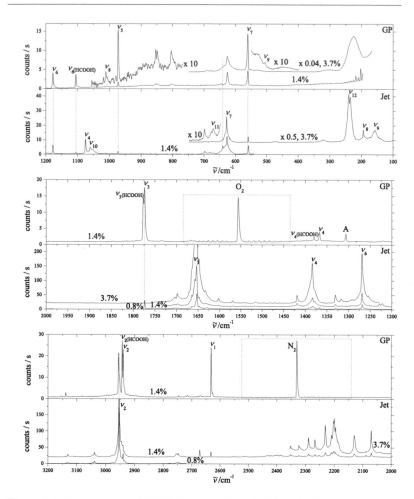

Figure 3.9: Raman spectra of HCOOD between 100 and 3200 cm^{-1}. The jet spectra were generally measured under three different conditions: a), $T_s = 16°C$ (3.7% HCOOD in He), $p_s = 700$ mbar, $d = 0.4$ mm, 8×200 s (for the spectrum between 100 and 700 cm^{-1}, $p_s = 500$ mbar, 12×200 s); b), $T_s = 0°C$ (1.4% HCOOD in He), $p_s = 700$ mbar, $d = 0.4$ mm, 8×100-200 s; c), $T_s = -8°C$ (0.8% HCOOD in He), $p_s = 700$ mbar, $d = 0.4$ mm, 8×100 s. The GP spectra were taken under the following conditions: $T_s = 0°C$ (1.4% HCOOD in He), $p_b = 30$ mbar, 4×60 s. Another GP spectrum in the low wavenumber region is shown for comparison: $T_s = 16°C$ (3.7% HCOOD in He), 6×30 s.

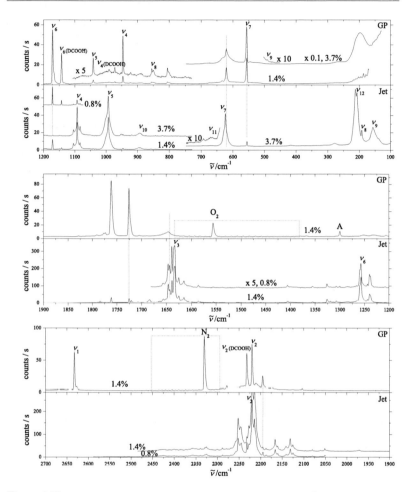

Figure 3.10: Raman spectra of DCOOD between 100 and 2700 cm^{-1}. The jet spectra were generally measured under three different conditions: a), $T_s = 16°C$ (3.7% DCOOD in He), $p_s = 700$ mbar, $d = 0.4$ mm, 8×200 s (for the spectrum between 100 and 750 cm^{-1}, $p_s = 500$ mbar, 12×200 s); b), $T_s = 0°C$ (1.4% DCOOD in He), $p_s = 700$ mbar, $d = 0.4$ mm, 8×200 s; c), $T_s = -8°C$ (0.8% DCOOD in He), $p_s = 700$ mbar, $d = 0.4$ mm, 8×100-150 s. The GP spectra were taken under the following conditions: $T_s = 0°C$ (1.4% DCOOD in He), $p_b = 30$ mbar, 4×60 s. Another GP spectrum in the low wavenumber region is shown for comparison: $T_s = 16°C$ (3.7% DCOOD in He), 6×30 s.

53

Mode	GP			Ar matrix	Calculation
	This work	IR	Raman[a]	IR[b]	
DCOOH					
ν_1	...	3570^c	3566	3551.7/3549.9	3664.6 (0.1)
ν_2	2220 (723)	2220^c	2218	2225.2	2307.5 (796.1)
ν_3	$1763/1726^d$	1756^c	$1760/1724^d$	1759.5/1722.9^d	1791.2 (31.7)
ν_4	972 (408)	971^e	...	976.2/974.1	990.2 (418.4)
ν_5	1206 (14)	1220^c	1297	1200.3/1199.6	1294.9 (4.7)
ν_6	1143 (−38)	1143^c	1140	1142.1/1141.7	1172.9 (−36.1)
ν_7	621 (5)	629^c	620	623.5/622.3	616.7 (5.8)
ν_8	850/854	870^c	...	875.0/874.8	879.9 (166.7)
ν_9	634 (8)	629^c	...	626.4/625.9	680.7 (9.8)
HCOOD					
ν_1	2632 (938)	2632^c, 2633.5^f	2631.4	...	2665.1 (999.6)
ν_2	2939 (4)	2948^c, 2944^f	$2938.2/2941.8^g$	...	3104.2 (−0.6)
ν_3	1774 (3)	1772^c, 1773^f	1773	...	1817.1 (5.8)
ν_4	1367 (13)	1360^c	1368	...	1399.0 (9.6)
ν_5	972 (248)	973^e, 972^f	972	...	1003.2 (296.4)
ν_6	1177 (−72)	1178^e, 1177.7^f	1177.9	...	1198.3 (−61.5)
ν_7	560 (−66)	541^c, 556.3^f	560.2	...	555.7 (−66.8)
ν_8	1011	1007^c, 1010.8^f	...	...	1043.8 (2.8)
ν_9	506 (136)	512^c, 507.5^f	...	...	543.4 (147.1)
DCOOD					
ν_1	2633 (937)	2632^c	2631.9^h	...	2665.6 (999.1)
ν_2	2232 (711)	2232^c	2231.8^h	...	2306.4 (797.2)
ν_3	$1761/1725^d$	1742^c	$1760.0/1723.5^{d,h}$	...	1789.4 (33.5)
ν_4	946 (434)	946^e	945.0^h	...	963.8 (444.8)
ν_5	1041 (179)	1040^e	1042^h	...	1053.9 (245.7)
ν_6	1171 (−66)	1171^e	1169.7^h	...	1192.3 (−55.5)
ν_7	556 (70)	538^c, 558^i	555.5^h	...	551.7 (70.8)
ν_8	849/855	873^c	...	...	879.7 (166.9)
ν_9	492 (150)	489^c, 491^i	...	...	525.8 (164.7)

[a] Bertie et al., Ref. 67
[b] Maçôas et al., Ref. 68
[c] Millikan and Pitzer, Ref. 70
[d] Fermi resonance doublet, see text for details.
[e] Williams, Ref. 10
[f] Hisatsune and Heicklen, Ref. 71
[g] Doublet band structure, see text for detailed discussion.
[h] Bertie and Michaelian, Ref. 9
[i] Miyazawa and Pitzer, Ref. 18

Table 3.4: Band maxima (in cm^{-1}) and assignment of the fundamental modes of the deuterated *trans*-FA, compared with the harmonic calculation at the B3LYP/6-31+G* level. The isotope redshifts (in cm^{-1}) relative to HCOOH observed in this work are listed in parentheses.

were assigned as $\nu_2/(\nu_3 + \nu_6)$ $(1774 + 1177 = 2951\,\mathrm{cm}^{-1})$. We prefer to assign the band a $2943\,\mathrm{cm}^{-1}$ to the C-H stretching vibration of HCOOH because this band is only observed in the GP spectrum but not in the HCOOD jet spectra, not like the weaker band at $2939\,\mathrm{cm}^{-1}$. O-H/D isotope exchange is much more effective in the GP measurements.

A large uncertainty exists for the C=O stretching vibration mode ν_3 of DCOOH /DCOOD. Doublet structures were observed in the spectra of both isotopomers: 1726 and $1763\,\mathrm{cm}^{-1}$ for DCOOH and 1725 and $1761\,\mathrm{cm}^{-1}$ for DCOOD. According to the interpretation from Bertie *et al.*, for DCOOH, the carboxyl stretching fundamental is in Fermi resonance with the combination band of $\nu_6 + \nu_7$ $(1143 + 621 = 1764\,\mathrm{cm}^{-1})$, yielding two bands of comparable intensity [9], whereas the related resonance partners for DCOOD are the ν_3 fundamental and the overtone band of the C-H out-of-plane bending mode ν_8 $(2 \times 873\,[10] = 1746\,\mathrm{cm}^{-1})$ [67]. In the Ar matrix work, the doublet of DCOOH was assigned as components of a Fermi resonance between ν_3 fundamental and the overtone band of ν_8 [68]. However, the ν_8 fundamental of both compounds is lower than $855\,\mathrm{cm}^{-1}$ according to our measurements, and the overtone bands should not be higher than $1710\,\mathrm{cm}^{-1}$. Compared with the doublet band positions at 1725 and $1761\,\mathrm{cm}^{-1}$ in the spectra of DCOOD, the combination band of $\nu_6 + \nu_7$ $(1171 + 556 = 1727\,\mathrm{cm}^{-1})$ is a better resonance partner for ν_3. However, the band positions of ν_3 of DCOOH and DCOOD is nearly the same, according to the calculation (see Tab. 3.4). It is hard to accept the coincidence that two observed bands will be at nearly the same positions after the resonance of ν_3 with another band having obviously different wavenumber.

According to the high resolution IR studies on the carboxyl stretch vibration of monomer DCOOH/DCOOD [76,77], The band center of ν_3 is determined to be $1725.8\,\mathrm{cm}^{-1}$ for DCOOH [77] and $1725.1\,\mathrm{cm}^{-1}$ for DCOOD [76]. The ν_3 band is strongly perturbed by the Fermi resonance with the $2\nu_8$ overtone band and Coriolis coupling to the combination band $\nu_6 + \nu_7$. The $2\nu_8$ overtone bands of both compounds have nearly the same band positions.

Furthermore, some unassigned bands will be discussed. The relatively strong band at $637\,\mathrm{cm}^{-1}$ in the jet spectra of DCOOH is assigned to the OCO bending mode of the mixed dimer of DCOOH and DCOOD [4], but it was noted that the band is unusually strong, indicating that there might be another contribution underneath. The ν_9 band of DCOOH monomer observed at $634\,\mathrm{cm}^{-1}$ in the GP spectra (see Fig. 3.8) possibly plays a role. It is relatively hard to imagine that such a large amount of DCOOD is contained in the DCOOH compound used in our measurement but a band with the same wavenumber was also observed in the GP spectra of DCOOD. Therefore, this significant impurity may come from the production process of the compounds (95% chemical purity). The band at $557\,\mathrm{cm}^{-1}$ in the GP spectra of DCOOH could also be from a chemical impurity.

There is a band at $644\,\mathrm{cm}^{-1}$ in the GP spectrum of HCOOD, on the right side of ν_7 of HCOOD monomer. It may correspond to the OCO bending mode of the mixed dimer of HCOOH and HCOOD at $641\,\mathrm{cm}^{-1}$ [4], which is clearly seen in the jet spectra at low acid concentration (1.4%, see Fig. 3.9).

Finally, there are some other bands in the GP spectra, which may be from the overtone/combination bands of monomers or chemical impurities. The assignment for them is very uncertain and therefore not discussed here.

3.1.3 Conclusions and outlook

The monomer fundamental modes of the four formic acid isotopomers are discussed. Comparison of the experimental values with quantum chemical calculations is made and only moderate agreement is found, which leads to difficulties in the assignment of several relative uncertain bands, e.g., ν_5 and ν_8.

No solid evidence for the existence of the energetically less stable *cis*-FA monomer was found. The band at $864\,\mathrm{cm}^{-1}$ observed in the Raman GP spectrum and assigned as the $\gamma(\mathrm{O-H})$ vibration of FFb [63] is unlikely to come from the monomer itself, but a joint assignment to a mixed complex of acid with water cannot be ruled out at this stage and a systematic investigation (measurement/calculation) on the mixed formic acid-water complex is needed.

3.2 Formic acid dimer

According to the analysis in the last section, the only kind of dimer conformer in the Raman jet spectra of formic acid in this work is FFa, which is a distinctly planar complex [78] with hydrogen bond-mediated restoring forces towards all kinds of tilting [3]. It involves periodic chemical bond breaking and making on the nanosecond timescale even at $0\,K$, due to facile concerted hydrogen transfer through high barriers along the equivalent hydrogen bonds [79]. These features come with the rare property that this dimer can be prepared and studied in the room temperature gas phase at high abundance.

Therefore, as the smallest organic complex with a double hydrogen bond but void of peripheral structural complexity, formic acid dimer (FAD) has been well studied by a range of experimental techniques [78, 80–82], in particular by vibrational spectroscopy [83]. The centro-symmetry turns infrared and Raman techniques into perfectly complementary approaches. Structural studies by microwave spectroscopy are difficult due to the inversion symmetry, but alternatives exist [84,85]. In view of the elementary character of FAD, it is not surprising that the number of theoretical approaches outnumbers the experimental ones (see, e.g., Refs. 86–91 and references cited therein). Some of these theoretical studies carry the quantum dynamical description to fairly accurate and high-dimensional levels [92 94], thus allowing for rigorous comparison to spectroscopic data.

Despite these intense experimental and theoretical studies, important experimental data are still missing. Such gaps slow down theoretical progress. Among them, the lack of anharmonicity information in the hydrogen bond modes is quite prominent [11,95]. The spectroscopy of vibrational fundamentals itself is already challenging, with the last missing van der Waals mode eluding discovery until 2007 [3]. The direct study of such van der Waals modes can supply very sensitive information on the hydrogen bond dynamics [96,97]. However, fundamental excitations only provide insights into the shape of the potential energy hypersurface close to the minimum structure, which is furthermore indirect because of anharmonicity. Larger excursions can be realized by overtone and combination transitions [11, 95, 98], which are not easily accessible in a system without low-lying electronic states [99]. Action spectra have nevertheless been obtained using selective photodissociation [98] and have yielded anharmonic coupling constants. However, these anharmonic studies concentrate on intramolecular modes. An indirect approach to hydrogen bond modes via combinations with hydride stretching vibrations [100] fails in this case, because carboxylic acid dimers exhibit a very complex dynamics in these hydrogen-bonded stretching states [3, 101]. The spectroscopy of higher-lying minima typically probes small excursions from those, rather than large excursions from the global minimum [63, 102, 103]. Both aspects are important for the hydrogen bond dissociation kinetics [104] and in the condensed phase [55, 82], but for a reliable force field of cyclic FAD, information in between is currently most needed.

FAD has 24 fundamental modes which may be classified according to their C_{2h} point group symmetry behavior and which correspond to 13 different kinds of vibrational motion. While the intermolecular twisting mode ν_{16} (in Herzberg notation [8], adopting, however, the representation sequence A_g, B_g, A_u, B_u [9]) is only IR active and the intermolecular stretching mode ν_8 is only Raman active, the other 11 vibrations occur in Davydov pairs of IR active and Raman active modes, which are split in a mode- and isotope-dependent way [67].

Many of them have been characterized a long time ago in the gas phase for dimers close to room temperature in thermodynamic equilibrium with the monomers. Infrared [18, 75] and Raman spectroscopy [9, 67] has been applied and isotope substitution was used to assist the assignments. The out-of-plane bending potential has occasionally been severely underestimated [105, 106], but apart from some assignment gaps, a satisfactory description of the harmonic force field has emerged towards the end of the last century. Nevertheless, the thermal excitation in regular gas phase spectra results in significant band shape distortions, because much less than 10% of the molecules are in the vibrational ground state at room temperature. This makes an accurate estimate of the band center difficult, unless high resolution spectroscopy reveals the hot band structure [11, 85, 107]. The latter is difficult in the Raman case and at low wavenumber in the infrared.

One way of reducing thermal excitation is the application of matrix isolation techniques [13–15]. It also gives access to less stable isomers, but may suffer from site splittings. Very recently, matrix isolation also became possible for Raman probing [69]. The spectral simplification offered by this technique comes at the price of matrix shifts, which are sometimes difficult to predict and can lead to changes in resonance patterns [15].

Therefore, more effort has been invested in infrared supersonic jet approaches over the last decade. These range from low resolution studies [101] using mostly cavity ring down techniques [95,108,109] over medium resolution FTIR spectra [11] and action spectra [98] to very high resolution analyses including tunneling splittings [79, 103, 110, 111]. Typically, the focus was on the C–H/O–H/C=O/C–O valence vibrations and on some of their isotopic variants. For the intermolecular modes in the far infrared range, no jet spectrum has been reported so far [11], although this would be technically possible [16, 96]. Very recently in our group, FIR measurements in supersonic jet have been realized and one of the three IR active intermolecular fundamentals of FAD has been observed (see Fig. 3.14 later on). Nevertheless, the Raman active half of the fundamentals is only beginning to be explored in the jet-cooled regime [3].

Using spontaneous Raman scattering combined with supersonic expansions, accurate band positions of the Raman active fundamentals are determined. Comparison of the experimental values, both Raman and IR, to harmonic/anharmonic quantum mechanical calculations is made to find the suitable calculation levels, which can suggest necessary corrections for the band positions of some IR active fundamentals, the low-temperature spectra data of which are missing. After this essential step the anharmonicity analysis becomes possible, from lower to higher wavenumber region, with more and more complicated combination mechanisms, following the way to fully characterize the low temperature vibrational dynamics of FAD.

An overview of the band positions of the FFa Raman active fundamental modes (see Figs. 3.2, 3.8, 3.9 and 3.10) is listed in Tab. 3.5, compared with the previous GP experiment [9,67]. The band positions are in good agreement, with band shifts

Description	Mode	(HCOOH)₂ Jet[a]	(HCOOH)₂ GP[b]	(DCOOH)₂ Jet[a]	(DCOOH)₂ GP[c]	(HCOOD)₂ Jet[a]	(HCOOD)₂ GP[c]	(DCOOD)₂ Jet[a]	(DCOOD)₂ GP[b]
A_g									
ν(O–H)	ν_1	...	...	...	...	2200	...	2215	...
ν(C–H)	ν_2	2952	2948.9	2212	2208	2954	2951.4	2220	2210.8
ν(C=O)	ν_3	1668	1669.8	1643	1643	1653	1679/1663	1633	1647.6
δ(O–H)	ν_4	1431	1415.0	1418	1385	1076	972	1093	1080.7
δ(C–H)	ν_5	1376	1374.8	1001	994	1382	1383.3	993	989.6
ν(C–O)	ν_6	1224	1214.0	1238	1230	1268	1260.9	1258	1250.0
δ(OCO)	ν_7	682	677.3	676	672	628	624	624	617.4
ν(O–H...O)	ν_8	194	...	193	...	194	...	192	...
δ(O–H...O)	ν_9	161	137.1	159	~140	157	~144	157	130
B_g									
γ(C–H)	ν_{10}	1055/1062	1059.7	891/897	...	1053/1060	1060	891/897	892
γ(O–H)	ν_{11}	907/915	...	918/924	...	671/677	...	665/670	...
γ(O–H...O)	ν_{12}	239/246	229.6	210/215	202	235/240	224	207/212	194.4

[a] Supersonic jet band centers, this work.
[b] Bertie and Michaelian, Raman GP spectra at room temperature (Ref. 9).
[c] Bertie et al., Raman GP spectra at room temperature (Ref. 67).

Table 3.5: Band positions (in cm⁻¹) and assignment of the Raman active fundamental modes of (HCOOH)₂ and its isotopomers.

within $10\,\mathrm{cm}^{-1}$. Exceptions are the intermolecular modes below $250\,\mathrm{cm}^{-1}$ and the O–H in-plane bending mode ν_4. It is very likely that the band positions of these modes are particularly sensitive to the vibrational temperature. All prominent jet-cooled bands with B_g symmetry show a doublet band structure with wavenumber differences between 5 to $7\,\mathrm{cm}^{-1}$, due to the rotational sub-branches with an expected rotational temperature of about $50\,\mathrm{K}$ [3, 4]. ν_{12} was chosen as an example to be discussed in detail later.

For two modes ν_1 and ν_2, the term values G as a function of quantum numbers v_1 and v_2 may be written as [8]:

$$
\begin{aligned}
G(v_1,\,v_2) \;=\; & \omega_1(v_1 + \tfrac{1}{2}) + \omega_2(v_2 + \tfrac{1}{2}) + x_{1,1}(v_1 + \tfrac{1}{2})^2 \\
& + x_{2,2}(v_2 + \tfrac{1}{2})^2 + x_{1,2}(v_1 + \tfrac{1}{2})(v_2 + \tfrac{1}{2}).
\end{aligned}
\tag{3.1}
$$

It follows that

$$
\nu_1 = G(1,\,0) - G(0,\,0) = \omega_1 + 2x_{1,1} + \tfrac{1}{2}x_{1,2},
\tag{3.2}
$$

$$
2\nu_1 = G(2,\,0) - G(0,\,0) = 2\omega_1 + 6x_{1,1} + x_{1,2},
\tag{3.3}
$$

$$
\nu_1 + \nu_2 = G(1,\,1) - G(0,\,0) = \omega_1 + \omega_2 + 2x_{1,1} + 2x_{2,2} + 2x_{1,2},
\tag{3.4}
$$

and the relevant anharmonicity constants $x_{1,1}$ and $x_{1,2}$ can be extracted from

$$
x_{1,1} = \tfrac{1}{2}(2\nu_1 - \nu_1 \cdot 2),
\tag{3.5}
$$

$$
x_{1,2} = (\nu_1 + \nu_2) - \nu_1 - \nu_2.
\tag{3.6}
$$

The O–H stretching dynamics turned out to be particularly challenging [11, 95, 109] due to interactions with combination bands from the lower frequency range. It has been known for several decades that vibrational spectra of the O–H and

N–H stretching bands show drastic changes upon H bonding: large redshifts and broadening which can amount to $500\,\text{cm}^{-1}$. For $(HCOOH)_2$, the structured bands spread over a range of about 2500 to $3200\,\text{cm}^{-1}$, whereas the bands of $(DCOOD)_2$ from 2000 to $2500\,\text{cm}^{-1}$ are even more complicated with interaction with C–D stretching modes. Such broadening is often accompanied with rich substructures, which suggests the presence of anharmonic coupling. Mechanisms such as Davydov splitting, Franck-Condon combinations with low-frequency hydrogen-bond modes, multiple Fermi resonances, hot bands, exchange tunneling, predissociation, or the breakdown of the Born-Oppenheimer approximation have been proposed.

Therefore, the overtone/combination bands of FFa will be discussed in three wavenumber regions respectively: the low-frequency intermolecular vibration region ($\leq 750\,\text{cm}^{-1}$), the finger-print intramolecular vibration region (750-$1800\,\text{cm}^{-1}$) and the O–H/D as well as C–H/D vibration region ($\geq 1800\,\text{cm}^{-1}$).

3.2.1 The region up to $750\,\text{cm}^{-1}$

A bottom-up approach which starts with a combined IR/Raman characterization of the lowest frequency modes, where regular dynamics prevails, and systematically progresses towards higher state densities appears indispensable for an indepth understanding of the complex and irregular hydride stretch dynamics. To minimize ambiguities, it is essential to know the extent of anharmonicity expected in the combination bands. This extent is particularly uncertain for the large amplitude van der Waals modes.

Fortunately, all six intermolecular modes fall below $300\,\text{cm}^{-1}$ [3, 9] and thus all their two-quantum overtones and combination bands are well separated from the lowest intramolecular modes, which start above $600\,\text{cm}^{-1}$. The IR coverage of unperturbed intermolecular fundamentals is limited to room temperature gas phase studies [11, 18, 106, 112, 113], whereas the Raman active fundamentals have been studied in the jet [3] and were recently confirmed in matrix isolation [69].

The interest in intermolecular combination bands is amplified by several quantum studies including anharmonic contributions [90, 114]. These studies indicate surprisingly small anharmonic constants, typically well below the accuracy limit of room temperature gas phase band centers [9, 11]. Therefore, an experimental investigation of some of these excited vibrational states appears particularly timely. Four weak Raman active overtones in this wavenumber region are unambiguously assigned with the help of systematic isotope substitution. In addition, a further overtone and very weak intermolecular combination bands are tentatively assigned. Furthermore, some previously observed IR-active intermolecular fundamental and combination bands of FAD are analyzed. Finally, the lowest intramolecular bands between 600 and $750\,\mathrm{cm}^{-1}$ are characterized and a weak fundamental is tentatively assigned a band center for the first time.

3.2.1.1 Experimental spectra

The cyclic formic acid dimer has six intermolecular vibrational modes (see Fig. 3.11), three of which are IR active. The lowest one, ν_{16} (A_u) near $69\,\mathrm{cm}^{-1}$, corresponds to the intermolecular twisting vibration. The others represent in-plane (ν_{24}, B_u) as well as out-of-plane bending motions (ν_{15}, A_u) [9]. Their fundamental wavenumbers have been reported in early low resolution IR gas phase work [18, 106, 112, 113] and were recently re-investigated [11]. In this context, the band center of ν_{15} was located reliably at $168.47\,\mathrm{cm}^{-1}$ using high resolution spectroscopy [11] whereas a substantial discrepancy was observed for ν_{24} (248-$268\,\mathrm{cm}^{-1}$ [11, 105, 113], *vide infra*). IR active combination bands have also been found early on [105, 113] and were summarized and assigned in Ref. 9. A combination band previously reported [105, 106] at $307\,\mathrm{cm}^{-1}$ was reallocated [11] at $311\,\mathrm{cm}^{-1}$. Apart from a lack of jet-cooled data and some inconsistencies in the low resolution band positions, a satisfactory knowledge of the IR-active hydrogen bond mode manifold may thus be diagnosed. However, it typically does not reach a level at which anharmonicity effects in these modes may be derived in a reliable way. We will come back to this issue in the following.

The other three intermolecular modes are Raman active. They represent inter-monomer stretching motion ν_8 (A_g) and symmetric in-plane ν_9 (A_g) as well as out-of-plane bending vibrations ν_{12} (B_g) [9]. The jet-cooled fundamental band

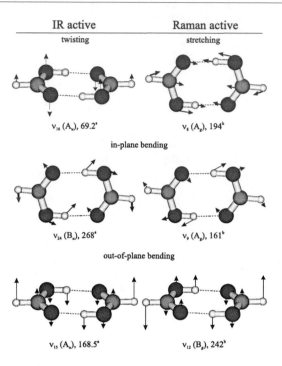

IR active Raman active

twisting stretching

ν_{16} (A$_u$), 69.2[a] ν_8 (A$_g$), 194[b]

in-plane bending

ν_{24} (B$_u$), 268[a] ν_9 (A$_g$), 161[b]

out-of-plane bending

ν_{15} (A$_u$), 168.5[a] ν_{12} (B$_g$), 242[b]

[a] Georges *et al.*, high resolution FIR spectrum in the gas phase, Ref. 11.
[b] Raman supersonic jet spectrum, this work.

Figure 3.11: Schematic drawings and experimental wavenumbers (in cm^{-1}) of the six inter-monomer fundamentals in the low frequency region.

positions have already been reported in a previous article [3] and were recently confirmed by matrix isolation Raman spectroscopy [69] with moderate matrix shifts. These results are summarized in Tab. 3.6 for comparison with the band positions of the present re-investigation shown in Fig. 3.12. The 67 L reservoir and the $8.0 \times 0.05\,\text{mm}^2$ nozzle were used for the measurements in Fig. 3.12. The chamber was evacuated by a $250\,\text{m}^3/\text{h}$ Roots pump backed by a $100\,\text{m}^3/\text{h}$ rotary vane pump.

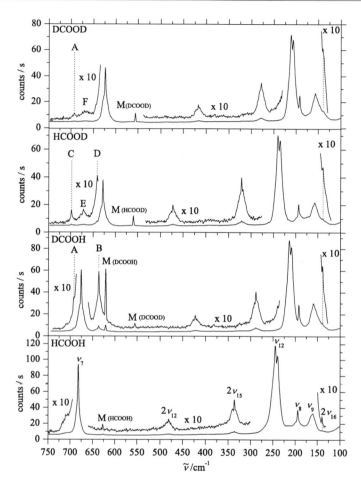

Figure 3.12: Raman jet spectra for the four H/D isotopomers of formic acid between 100 and 750 cm^{-1} under dimer-dominant expansion conditions. For all the four acids, $T_s = 16°$C (3.7% acid in He), $d = 0.4$ mm. For HCOOH, $p_s = 1000$ mbar, 12×300 s; for the other three isotopomers, $p_s = 500$ mbar, 12×200 s (12×240 s for DCOOH). The labels A-F refer to mixed isotopomer and intramolecular torsion assignments listed in Tab. 3.11. See text for details.

In addition, four weakly Raman active overtone transitions are assigned based on Fig. 3.12 and their band maxima are also listed in Tab. 3.6. Only the overtone of the IR-active vibration ν_{15} had been previously reported in the Raman gas phase spectra [9, 67], 5% below the jet-cooled position. The 5% red shift may be explained by thermal weakening of the hydrogen bond at room temperature. A very weak band at $598.8 \pm 1.9\,\mathrm{cm}^{-1}$ mentioned for $(HCOOH)_2$ in the room temperature Raman gas phase spectrum [9] could not be confirmed in the present work. Other Raman active overtone/combination bands have not been reported in the literature.

We shall start with the discussion of slight wavenumber differences between the earlier jet work [3] and the present one in Tab. 3.6. Most differences are within $1\,\mathrm{cm}^{-1}$ and may be explained by integer rounding errors, calibration uncertainties and slight shifts due to cluster contributions beyond the dimer. The only systematic differences of $1\text{-}4\,\mathrm{cm}^{-1}$ refer to the lowest frequency mode in the previous study, the in-plane bending fundamental ν_9. Here, the earlier determination of the

$(HCOOH)_2$			$(DCOOH)_2$		$(HCOOD)_2$		$(DCOOD)_2$		
Jet[a]	Jet[b]	Matr.[c]	Jet[a]	Jet[b]	Jet[a]	Jet[b]	Jet[a]	Jet[b]	Assignment
139	...	...	139 (0)	...	139 (0)	...	139 (0)	...	$2\nu_{16}$
161	165	165	159 (2)	162	157 (4)	160	157 (4)	158	ν_9
194	194	196	193 (1)	192	194 (0)	194	192 (2)	192	ν_8
242	242	251[d]	212 (30)	212[e]	238 (4)	237	210 (32)	209	ν_{12}
319?	...	...	...	...	...	...	...	...	$2\nu_9$
336	...	...	289 (47)	...	321 (15)	...	278 (58)	...	$2\nu_{15}$
386	...	...	...	...	...	...	...	...	$2\nu_8$
400	...	...	...	...	...	...	...	...	$\nu_{12} + \nu_9$
435?	...	...	...	...	...	...	...	...	$\nu_{12} + \nu_8$
482	...	...	423 (59)	...	473 (9)	...	417 (65)	...	$2\nu_{12}$

[a] This work.
[b] Zielke and Suhm, Ref. 3.
[c] Olbert-Majkut *et al.*, Ref. 69 (Argon matrix, 19 K).
[d] A band at $243\,\mathrm{cm}^{-1}$ was also assigned to ν_{12}, see text for details.
[e] Typographical error in Ref. 3 (222 instead of $212\,\mathrm{cm}^{-1}$), see also Ref. 33.

Table 3.6: Band maxima (in cm^{-1}) and assignment of Raman active low frequency modes of $(HCOOH)_2$ and its isotopomers. The isotope red shifts (in cm^{-1}) relative to $(HCOOH)_2$ observed in the present work are listed in parentheses.

band maximum was affected by a spectral drift in the holographic notch filter used in Ref. 3 (Kaiser Optical Systems, $\varnothing = 62\,\text{mm}$, 532.0-2.5 nm) and the associated efficiency drop upon aging. In the present work, a more stable Raman edge filter with cutoff closer to the Rayleigh line was employed and allows for a more reliable determination of the band maximum in this broad band in Fig. 3.12.

When comparing the present jet data to the recent matrix isolation spectra, the ν_{12} fundamental deserves detailed discussion. In the jet spectra, the quoted band center estimate is actually midway between two peaks separated by $\approx 5\,\text{cm}^{-1}$. The symmetry of the vibration, the similar shape and height of the two components and the invariance with respect to isotope substitution in all three cases led us to an assignment of these two subbands as $\Delta J \neq 0$ transitions. Their separation is consistent with an expected rotational temperature of $\approx 50\,\text{K}$ at a distance of ≈ 8 nozzle diameters. This is significantly higher than the temperature obtained in an IR cavity-ring-down study [108], because the expansion is probed so close to the nozzle exit for the Raman spectrum. With the improved signal-to-noise ratio of the present spectra, we have no reason to question this original assignment [3], although a band shape simulation using Raman intensities and rigid rotor rotational constants would be desirable. Surprisingly, the matrix isolation spectra also contain two bands [69], although rotation is certainly suppressed in the Ar host. One band matches the jet value quite closely, whereas the other, stronger one is $8\,\text{cm}^{-1}$ higher in wavenumber. This could be a site-splitting phenomenon in the Ar matrix [15] or less likely due to Raman activation of an IR-active mode by a symmetry-breaking effect. The former effect would help to explain the detuning of a gas phase Fermi resonance by the rare gas surrounding [15]. The authors of the Raman matrix study also consider the possibility of a combination band $\nu_{16} + \nu_{15}$ for the origin of the *higher* frequency component. Based on IR evidence for the component bands and their hot band transitions [11], we would argue that this is unlikely. A combination band origin of the *lower* transition is more conceivable. In the jet spectra, this combination band is predicted near $236.5\,\text{cm}^{-1}$ and thus on the low wavenumber wing of ν_{12}. Note that for symmetry reasons neither this nor any other state in the vicinity can profit from Fermi resonance with ν_{12} (B_g).

To analyze this Ar-matrix effect further, we have carried out supersonic jet expansions in Ar at different nozzle distances. Previous work [53] has shown that this leads to amorphous layers of Ar around the molecules. Fig. 3.13 shows the result in the hydrogen bond fundamental range (upper figure), compared with that of the first intramolecular fundamentl ν_7 (lower figure). All Raman-active bands experience blue shifts which in some cases even exceed those in the bulk Ar matrix [69]. This shows that the hydrogen bond modes of FAD are hindered by the matrix environment. The shift of the ν_{12} band is particularly pronounced, in qualitative agreement with the expectation for an intermolecular out-of-plane bending mode. For comparison, the intramolecular ν_7 mode does not show a significant shift. The survival of a second peak at $245\,\mathrm{cm}^{-1}$ besides the dominant ν_{12} band at $257\,\mathrm{cm}^{-1}$ under strongly Ar-coating conditions (3 mm nozzle distance) indeed supports the existence of a second transition on its low-wavenumber wing. This might be the $\nu_{16} + \nu_{15}$ combination band proposed before [69], but its intensity would be unusually large.

Site-splittings are less likely for the amorphous Ar arrangements around the molecule. The substantial Ar shifts underscore the importance of vacuum-isolated spectra in the comparison of experimental and quantum-chemical data. Ar-nanocoating in the jet clearly offers a way to better understand bulk Ar matrix spectra. One may speculate that the new features in the Ar expansion arise from a polar dimer, which is not present in the He expansion. However, our current understanding of the Ar coating process is that it happens downstream of the acid dimerization and will not lead to isomerization. This is different from the He droplet behaviour [103].

As in the previous work [3], isotope substitution have been used to corroborate the mode assignment of the Raman active hydrogen bond fundamentals (see Tab. 3.6). The ν_9 mode shows a weak isotope dependence, which is also tabulated directly in parentheses. The isotope pattern of ν_8 is also weakly pronounced, but it contains a counterintuitive trend which has been explained in part by off-diagonal anharmonicity contributions [3]. The ν_{12} mode is particularly sensitive to C-deuteration, as one might expect from Fig. 3.11.

Intermolecular fundamentals

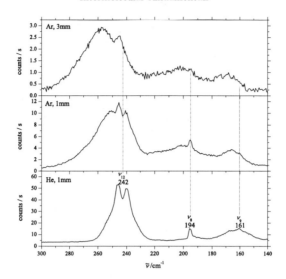

The OCO in-plane bending mode ν_7 (lowest intramolecular mode of FAD)

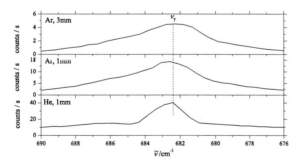

Figure 3.13: Effect of Ar-nanocoating on the Raman-active hydrogen bond fundamentals of FAD in expansions of of $\approx 1\%$ FAD in different carrier gases. All the spectra were measured with a $4.0 \times 0.15\,\mathrm{mm}^2$ slit nozzle, $p_\mathrm{s} = 700\,\mathrm{mbar}$. For both figures, bottom trace: $T_\mathrm{s} = -2°\mathrm{C}$ (1.2% HCOOH in He), $d = 1\,\mathrm{mm}$, $6 \times 200\,\mathrm{s}$; center trace: $T_\mathrm{s} = -5°\mathrm{C}$ (1.0% HCOOH in Ar), $d = 1\,\mathrm{mm}$, $10 \times 300\,\mathrm{s}$; top trace: $T_\mathrm{s} = -5°\mathrm{C}$ (1.0% HCOOH in Ar), $d = 3\,\mathrm{mm}$, $10 \times 300\,\mathrm{s}$.

3.2.1.2 Comparison with harmonic predictions

While the assignment of the three Raman-active fundamentals based on isotope substitution patterns is beyond doubt [3], it is instructive to compare the experimental anharmonic transitions with harmonic predictions at different levels of quantum-chemical approximation. Tab. 3.7 contains such a comparison using inexpensive methods and basis sets which leave room for a systematic sampling of the multidimensional potential energy hypersurface [92–94]. Already for the three Raman active fundamentals, it is obvious that the simple MP2/6-31+G* level shows the best agreement in this comparison. The deviations are on the order of 1% except for the intermolecular stretching mode, where the prediction is 5% too high. The latter finding is not too surprising, because isotope effects had revealed a substantial off-diagonal anharmonicity contribution in this mode [3], which has the effect of reducing the effective stretching force constant for light hydrogen bonds compared to heavier deuterium bonds. The proper comparison between experiment and theory at the force constant level would involve an infinitely heavy hydrogen atom and the true harmonic wavenumber of $(HCOOH)_2$ is thus likely

Mode	experiment	B3LYP aug-cc-pVTZ[a]	MP2 aug-cc-pVTZ[a]	6-31+G*	6-311+G*
ν_{16} (A_u)	$68^b,69.2^c$	76.1	69.3	68.8	63.7
ν_9 (A_g)	161^d	176.4	169.1	162.1	159.1
ν_{15} (A_u)	168.5^c	184.4	183.3	169.0	142.9
ν_8 (A_g)	194^d	212.2	212.3	202.9	191.6
ν_{12} (B_g)	242^d	260.2	260.0	242.3	209.0
ν_{24} (B_u)	$248^b,268^e,264^f$	282.8	281.2	253.7	237.6

[a] Zielke, Ref. 33, similar values were obtained before [88].
[b] FIR spectrum in the gas phase, Refs. 105,113.
[c] Georges et al., high resolution FIR spectrum in the gas phase, Ref. 11.
[d] Raman supersonic jet spectrum, this work.
[e] Georges et al., Ref. 11, see text for discussion.
[f] IR jet spectrum, courtesy of Kollipost.

Table 3.7: Comparison of calculated harmonic FAD low frequency fundamentals using different quantum chemical methods and basis sets with experimental (anharmonic) values (in cm^{-1}).

to be somewhat closer to $200\,\text{cm}^{-1}$ than to the experimental anharmonic fundamental of $194\,\text{cm}^{-1}$. Nevertheless, the double hydrogen bond is probably slightly weaker than predicted at the MP2/6-31+G* level.

At the B3LYP/aug-cc-pVTZ level, all Raman active fundamentals are systematically overestimated by up to 10%. This is also the case for the MP2/aug-cc-pVTZ level, whereas the MP2/6-311+G* calculations severely underestimate the stiffness of the FAD plane. By fortuitous error cancellation, the MP2/6-31+G* level may therefore be viewed as a surprisingly accurate zeroth order description of the Raman active hydrogen bond fundamentals. This remains true when the best available values for the IR active van der Waals modes are included in the analysis (see Tab. 3.7). Again, the B3LYP/aug-cc-pVTZ potential hypersurface appears to be too stiff, whereas the MP2/6-311+G* hypersurface is far too soft, in particular in the out-of-plane modes. The surprising fidelity of the MP2/6-31+G* description of the FAD fundamentals carries over to all isotopomers (Tab. 3.8) and even the largest relative error in the ν_8 mode decreases with deuteration, as it should. While we will not focus on Raman scattering strengths in this work, it may be mentioned that the predicted intensity ratio is about 1:10:50 for ν_8:ν_9:ν_{12}, in fairly good agreement with experiment.

There is IR information which requires further discussion. The ν_{24} fundamental was recently located at $268\,\text{cm}^{-1}$ [11], revising earlier assignments near $248\,\text{cm}^{-1}$ [105,113]. The earlier value fits the calculated harmonic MP2/6-31+G* prediction much better, whereas the new proposal is more in line with MP2/aug-cc-pVTZ predictions, which failed for the out-of-plane bending mode (Tab. 3.7). On the other hand, we note that the reduced transmission of the beam splitter caused excessive noise around $250\,\text{cm}^{-1}$ in Ref. 11 and may have complicated the assignment, although it is a rotationally resolved study. Therefore, the $248\,\text{cm}^{-1}$ was used as the band position of the ν_{24} fundamental for the analysis in our former study [4]. In that work we have also assumed that the $0\,\text{K}$ value of ν_{24} is likely to fall in between the two values.

This discrepancy can be resolved by the FIR jet spectra shown in Fig. 3.14. Only a single band at $264\,\text{cm}^{-1}$ was observed in the spectra under different con-

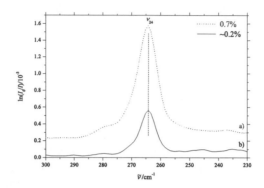

Figure 3.14: FTIR jet spectra of the ν_{24} fundamental of $(HCOOH)_2$ in He between 230 and 300 cm^{-1} at two different conditions, measured by F. Kollipost and R. Larsen. a), $T_s = -10°C$, $p_0 = 1.6$ bar, $p_s = 700$ mbar; b), $T_s = -15°C$ (1.4% HCOOH in He), $p_0 = \sim 3.0$ bar (two He flows with ~ 1.6 bar pressure were used as carrier gas to reduce the HCOOH concentration), $p_s = 700$ mbar.

ditions and it could be assigned straightforwardly as the ν_{24} fundamental. Our band position fits well to that from the high resolution GP study [11]. Their experimental vales of the other two IR active fundamentals below 200 cm^{-1} will be used for the further analysis.

At this stage, one must concede that the good agreement between experiment and MP2/6-31+G* calculations rests on a comparison between anharmonic and harmonic values, in spite of the relatively much overestimated band position of the ν_8 fundamental and the underestimated ν_{24} fundamental (see Tab. 3.7). Overtones and combination bands are needed to judge whether it also holds at the purely harmonic level. For this purpose, the 6-7 new weak bands observed in this study turn out to be valuable. Their isotope patterns (where available, see Tab. 3.6 and Fig. 3.12) and band positions suggest straightforward assignments. The lowest one must be the overtone of the IR-active twisting mode, $2\nu_{16}$. It coincides with potential absorptions from air impurities, but the absence of other rotational transitions of O_2 and N_2 in the vicinity rules out a major distortion of the band profile. The highest transition correlates nicely with twice the dominant Raman active fundamental wavenumber (Fig. 3.12, out-of-plane bending) and is thus assigned to $2\nu_{12}$. Its photon count is about 100 times lower than for the fundamental. The weakest sharp band (see Fig. 3.15 below) corresponds to the overtone of the stretching fundamental, $2\nu_8$. It scatters about 200 times less photons than the fundamental. The strongest non-fundamental transition has no consistent Raman active combination counterpart, but it matches the isotope pattern of the infrared out-of-plane bending fundamental and is thus assigned to $2\nu_{15}$.

This means that three out of the four Raman-active overtone transitions correspond to twisting or bending motions of FAD out of its planar minimum structure. Together with the corresponding fundamentals, these overtones of ν_{16}, ν_{15} and ν_{12} span a systematic data set for the characterization of anharmonicity in these distortions. Note that the overtone of the in-plane rocking mode ν_9, which shows some promise in promoting the hydrogen exchange tunneling [90], could not be safely identified in the regular Raman spectrum at the present signal-to-noise ratio. Its scattering strength must be a factor of 10 weaker than that of the neighboring $2\nu_{15}$ state, at least. As shown below, a polarization experiment can at least provide tentative evidence for it by removing the depolarized fraction of the $2\nu_{15}$ band.

Tab. 3.8 lists harmonic predictions for the proposed bending and stretching overtones for all four symmetric isotopomers, taking simply twice the calculated harmonic fundamental wavenumber. Isotope shifts are given in parentheses. They vary between nearly 0 and almost $70\,\mathrm{cm}^{-1}$. Both the observed band positions and the wavenumber shifts between FAD and its isotopically substituted counterparts in the present work match the calculated ones very well, rendering the assignments given in Tab. 3.6 straightforward.

The ν_{16} and ν_{12} overtones which we find in this work had not been assigned before. The overtone band of ν_{15} had been reported in previous Raman room temperature gas phase spectra, with wavenumbers of $317.7 \pm 1.4\,\mathrm{cm}^{-1}$ for (HCOOH)$_2$ [9], $275 \pm 2\,\mathrm{cm}^{-1}$ for (DCOOH)$_2$ [67], $305 \pm 3\,\mathrm{cm}^{-1}$ for (HCOOD)$_2$ [67], as well as $266.2 \pm 1.3\,\mathrm{cm}^{-1}$ for (DCOOD)$_2$ [9]. The systematic red shifts in the gas phase

Mode	$(HCOOH)_2$	$(DCOOH)_2$	$(HCOOD)_2$	$(DCOOD)_2$
$2\nu_{16}$ [a]	137.6	136.6 (1.0)	137.4 (0.2)	136.6 (1.0)
ν_9	162.1	161.3 (0.8)	158.3 (3.8)	157.7 (4.4)
ν_8	202.9	200.4 (2.5)	201.7 (1.2)	199.3 (3.6)
ν_{12}	242.3	212.2 (30.1)	236.9 (5.4)	208.5 (33.8)
$2\nu_{15}$ [a]	338.0	290.2 (47.8)	323.0 (15.0)	279.8 (58.2)
$2\nu_{12}$ [a]	484.6	424.4 (60.2)	473.8 (10.8)	417.0 (67.6)

[a] Twice the calculated fundamental wavenumbers.

Table 3.8: Calculated harmonic wavenumbers (in cm^{-1}) for FAD low frequency modes at the MP2/6-31+G* level. The calculated wavenumber shifts (in cm^{-1}) between $(HCOOH)_2$ and its three isotopomers are listed in parentheses.

spectra compared to the jet spectra are mostly due to thermal bond weakening effects in the former. This is also seen in the gas phase trace shown in Fig. 3.15. The thermal shifts amount to 4-5% and are capable of masking any anharmonicity effect in FAD. This clearly shows that jet-cooled spectra are indispensable for a straightforward anharmonicity analysis, as it will be carried out in the following.

3.2.1.3 Anharmonicity analysis

From the jet Raman overtones and fundamentals, one obtains for the ν_{12} mode of the different isotopomers an average diagonal anharmonicity constant $x_{12,12}$ of $-(1 \pm 1)\,cm^{-1}$ (see Tab. 3.9), according to Eq. 3.5. This persistently negative value is just slightly larger than the error bar due to calibration and residual band center uncertainties from the band profile. We conclude that a jet measurement is essential for its detection, because thermal shifts in the room temperature gas phase are an order of magnitude larger and matrix isolation shifts are difficult to predict.

For the ν_{15} mode, the infrared fundamental band center for the main isotopomer is available from a rotationally resolved measurement [11] in good agreement with earlier gas phase work [105, 113]. Therefore, an anharmonic analysis based on the Raman overtone is also possible and yields $x_{15,15} = -0.5\,cm^{-1}$. For the isotopomers, no reliable experimental IR fundamentals are available, but if we take the calculated harmonic fundamental band positions in Tab. 3.8 as an approximate reference validated by the HCOOH result, the $x_{15,15}$ are all slightly negative and probably do not exceed 1-2 cm^{-1} in magnitude (see Tab. 3.9).

Dimer	Mode	$x_{i,j}$ / cm^{-1}	
		anharmonic calculation	experiment
(HCOOH)$_2$	$2\nu_{16}$ (A$_g$)	$(141.6 - 71.3 \times 2) / 2 = -0.5$	$(139^a - 69.2^b \times 2) / 2 = \mathbf{0}$
	$2\nu_9$ (A$_g$)	$(326.5 - 163.9 \times 2) / 2 = -0.7$	$(319^a - 161^b \times 2) / 2 = \mathbf{-2}$
	$2\nu_{15}$ (A$_g$)	$(346.7 - 174.4 \times 2) / 2 = -1.1$	$(336^a - 168.5^b \times 2) / 2 = \mathbf{-1}$
	$2\nu_8$ (A$_g$)	$(376.5 - 190.2 \times 2) / 2 = -1.9$	$(386^a - 194^a \times 2) / 2 = \mathbf{-1}$
	$2\nu_{12}$ (A$_g$)	$(490.1 - 246.0 \cdot 2) / 2 = -1.0$	$(482^a - 242^a \times 2) / 2 = \mathbf{-1}$
	$\nu_{12} + \nu_9$ (B$_g$)	$407.9 - 246.0 - 163.9 = -2.0$	$400^a - 242^a - 161^a = \mathbf{-3}$
	$\nu_{12} + \nu_8$ (B$_g$)	$433.2 - 246.0 - 190.2 = -3.0$	$435?^a - 242^a - 194^a = \mathbf{-1}$
	$\nu_{12} + \nu_{16}$ (B$_u$)	$315.9 - 246.0 - 71.3 = -1.4$	$311^b - 242^a - 69.2^b = \mathbf{0}$
	$\nu_9 + \nu_{15}$ (A$_u$)	$336.4 - 163.9 - 174.4 = -1.9$	$(329^c - 161^a - 168.5^b = -1)$
	$\nu_{12} + \nu_{15}$ (B$_u$)	$417.3 - 246.0 - 174.4 = -3.1$	$(395^c - 242^a - 168.5^b = -16)$
	$\nu_9 + \nu_{24}$ (B$_u$)	$425.1 - 163.9 - 264.6 = -3.4$	$(395^c - 161^a - 264^e = -30)$
(DCOOH)$_2$	$2\nu_{16}$ (A$_g$)	$(141.5 - 71.3 \times 2) / 2 = -0.6$	$(139^a - 68.3^d \times 2) / 2 = \mathbf{1}$
	$2\nu_{15}$ (A$_g$)	$(297.2 - 149.2 \times 2) / 2 = -0.6$	$(289^a - 145.1^d \times 2) / 2 = \mathbf{-1}$
	$2\nu_{12}$ (A$_g$)	$(429.4 - 215.4 \times 2) / 2 = -0.7$	$(423^a - 212^a \times 2) / 2 = \mathbf{-1}$
(HCOOD)$_2$	$2\nu_{16}$ (A$_g$)	$(141.6 - 71.5 \times 2) / 2 = -0.7$	$(139^a - 68.7^d \times 2) / 2 = \mathbf{1}$
	$2\nu_{15}$ (A$_g$)	$(334.3 - 168.2 \times 2) / 2 = -1.1$	$(321^a - 161.5^d \times 2) / 2 = \mathbf{-1}$
	$2\nu_{12}$ (A$_g$)	$(484.7 - 243.5 \times 2) / 2 = -1.2$	$(473^a - 238^a \times 2) / 2 = \mathbf{-2}$
	$\nu_{12} + \nu_{16}$ (B$_u$)	$313.4 - 243.5 - 71.5 = -1.6$	$(299^c - 238^a - 68.7^d = -8)$
	$\nu_{12} + \nu_{15}$ (B$_u$)	$408.6 - 243.5 - 168.2 = -3.1$	$(390^c - 238^a - 161.5^d = -10)$
	$\nu_9 + \nu_{24}$ (B$_u$)	$413.3 - 160.6 - 256.0 = -3.3$	$(390^c - 157^a - 248.1^d = -15)$
(DCOOD)$_2$	$2\nu_{16}$ (A$_g$)	$(141.4 - 71.4 \times 2) / 2 = -0.7$	$(139^a - 68.3^d \times 2) / 2 = \mathbf{1}$
	$2\nu_{15}$ (A$_g$)	$(289.1 - 145.2 \times 2) / 2 = -0.7$	$(278^a - 139.9^d \times 2) / 2 = \mathbf{-1}$
	$2\nu_{12}$ (A$_g$)	$(425.5 - 213.5 \times 2) / 2 = -0.8$	$(417^a - 210^a \times 2) / 2 = \mathbf{-2}$
	$\nu_{12} + \nu_{16}$ (B$_u$)	$283.9 - 213.5 - 71.4 = -1.0$	$(277^e - 210^a - 68.3^d = -1)$

[a] Supersonic jet band centers, this work.
[b] Georges *et al.*, high resolution FIR spectrum in the gas phase [11].
[c] Carlson *et al.*, low resolution FIR gas phase spectrum [105].
[d] Calculated harmonic wavenumbers at the MP2/6-31+G* level, this work.
[e] IR jet spectrum, courtesy of Kollipost.
[f] Clague and Novak, low resolution FIR gas phase spectrum [113].

Table 3.9: Comparison of calculated B3LYP/6-311++G(2d,2p) anharmonic constants derived from intermolecular overtone and combination bands of FAD and its isotopomers with experimentally derived constants using experimental jet and gas phase data as well as theoretical isotope extrapolation. Rigorous experimental values are given in bold face, whereas combinations relying on unresolved gas phase bands are given in parentheses.

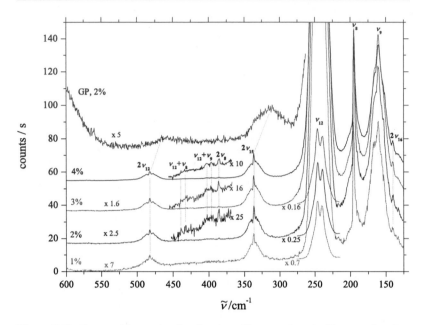

Figure 3.15: Raman jet spectra for formic acid expansions in He as a function of approximate concentration, scaled to similar scattering intensity. All the spectra were measured with a $4.0 \times 0.15 \, \text{mm}^2$ slit nozzle, $T_s = 20/15/9/-2°C$ from top to bottom, $p_s = 700 \, \text{mbar}$, $d = 0.4 \, \text{mm}$, $6 \times 200 \, \text{s}$. For the weaker bands, a 300 K GP spectrum is shown for comparison: 2.3% HCOOH in He $(T_s = 8°C)$, $p_b = 30 \, \text{mbar}$, $4 \times 30 \, \text{s}$.

The same is true for the very weak first overtone of the dimer stretching mode ν_8 (Fig. 3.15) which features an anharmonicity constant $x_{8,8} = -1 \, \text{cm}^{-1}$. The fact that this overtone band has a width of less than $2 \, \text{cm}^{-1}$ puts a loose upper bound on the hydrogen transfer tunneling splitting in the excited state. The assignment of $2\nu_8$ is further confirmed by a depolarization measurement, because the dimer stretching mode is predicted to give rise to a strongly polarized band. Fig. 3.16 shows a comparison of the spectrum recorded with perpendicular polarization of the excitation laser and the polarized component alone (bottom). The latter is estimated by subtracting the spectrum with parallel laser polarization, multiplied by 7/6, from the spectrum with perpendicular laser polarization. One can see that the band at $385\,\mathrm{cm}^{-1}$ persists quite strongly, as does ν_8. The persistence of signal at ν_{12} is in part a consequence of its strong scattering intensity which reacts sensitively to concentration variations, but may also reflect an overlapping A_g band, as speculated above. Furthermore, the polarized spectrum reveals a very weak band at $319\,\mathrm{cm}^{-1}$ (marked with an arrow), which we tentatively assign to $2\nu_9$, again consistent with a very small diagonal anharmonicity constant of $-1.5\,\mathrm{cm}^{-1}$. It is more than 200 times weaker than the fundamental.

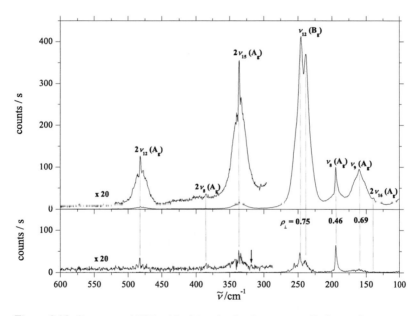

Figure 3.16: Spectrum of FAD with the excitation laser perpendicular to the scattering plane (top trace) and residual after subtracting 7/6 of the spectrum obtained with the excitation laser parallel to the scattering plane. For both spectra, $T_s = 20°C$ (4.5% HCOOH in He), $p_s = 700\,\mathrm{mbar}$, $d = 0.4\,\mathrm{mm}$, $12 \times 120\,\mathrm{s}$ and a $4.0 \times 0.15\,\mathrm{mm}^2$ slit nozzle was used. A_g bands with a small depolarization ratio $\rho_\perp$ persist most.

A larger uncertainty exists for the analysis of diagonal anharmonicity in ν_{16} because the IR fundamental has not been analyzed at rotational resolution [11] and the Raman overtone is weak and possibly affected by nearby air impurity absorptions. However, there is satisfactory agreement between different lower resolution studies of the fundamental [11, 105, 113]. Within these uncertainties, the anharmonicity constant $x_{16,16}$ is negligible.

Raman active combination bands are found to be extremely weak. Fig. 3.15 indicates evidence for $\nu_{12} + \nu_9$ and very circumstantial evidence for $\nu_{12} + \nu_8$ (see Tab. 3.9). If these assignments are correct, they support very small anharmonic cross terms.

The Raman insights obtained for the fundamentals, overtones and combination bands may be used to re-analyze some IR work in this spectral range. Three IR active van der Waals combination bands of FAD and some corresponding isotopomer bands were reported in earlier work [9, 11, 105, 113] (see Tab. 3.10 and

Experiment (in cm^{-1})	Dimer	Assignment		Prediction (in cm^{-1})
		Ref. 9	This work	
311^a, $307^{b,c}$	(HCOOH)$_2$	$\nu_{16} + \nu_{12}$	$\nu_{16} + \nu_{12}$	$68.8 + 242 = 310.8$
		$\nu_{15} + \nu_9$		$169.0 + 161 = 330.0$
299^b	(HCOOD)$_2$	$(\nu_{16} + \nu_{12})$	$\nu_{16} + \nu_{12}$	$68.7 + 238 = 306.7$
		$(\nu_{15} + \nu_9)$		$161.5 + 157 = 318.5$
277^c	(DCOOD)$_2$	$\nu_{16} + \nu_{12}$	$\nu_{16} + \nu_{12}$	$68.3 + 210 = 278.3$
		$\nu_{15} + \nu_9$		$139.9 + 157 = 296.9$
329^b	(HCOOH)$_2$		$\nu_{15} + \nu_9$	$169.0 + 161 = 330.0$
395^b	(HCOOH)$_2$	$\nu_{15} + \nu_{12}$	$\nu_{15} + \nu_{12}$	$169.0 + 242 = 411.0$
			$\nu_{24} + \nu_9$	$253.7 + 161 = 414.7$
390^b	(HCOOD)$_2$	$(\nu_{15} + \nu_{12})$	$\nu_{15} + \nu_{12}$	$161.5 + 238 = 399.5$
			$\nu_{24} + \nu_9$	$248.1 + 157 = 405.1$

[a] Georges et al., Ref. 11 (High resolution FIR gas phase spectrum).
[b] Carlson et al., Ref. 105 (FIR gas phase spectrum).
[c] Clague and Novak, Ref. 113 (FIR gas phase spectrum).

Table 3.10: Tentative assignment of some IR active combination bands of FAD and its isotopomers oberserved in earlier far infrared gas phase work [11, 105, 113]. Predictions are made based on the calculated band positions of IR active fundamental modes at the MP2/6-31+G* level (slanted) and the jet Raman data of this work, neglecting mixed anharmonicity contributions. Assignments in parentheses for (HCOOD)$_2$ were not proposed in Ref. 9.

Fig. 3.17). The lowest frequency one of $(HCOOH)_2$ was observed at $307\,cm^{-1}$ (Ref. 105, 113) to $311\,cm^{-1}$ (Ref. 11). Related band positions for $(HCOOD)_2$ $((DCOOD)_2)$ were reported [105] at $299\,cm^{-1}$ $(277\,cm^{-1}$ in Ref. 113). Two higher energy IR active combination bands at $329\,cm^{-1}$ and $395\,cm^{-1}$ were detected in Ref. 105. Bertie and Michaelian assigned the non- and fully deuterated bands in the room temperature gas phase spectra [9], neglecting anharmonic contributions.

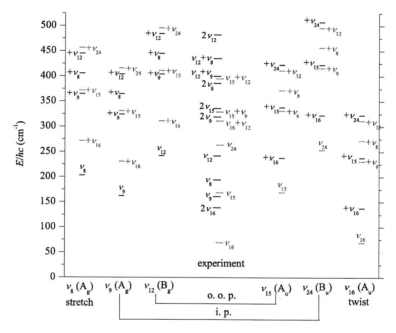

Figure 3.17: Comparison of the predicted and experimental band positions of FAD below $500\,cm^{-1}$ up to two vibrational quanta. The Raman active bands are marked with black solid lines while the IR active ones are shown as red dotted lines. The experimental data (center, Raman transitions from this work and Ref. 3, IR transitions from Ref. 11 (ν_{15} and ν_{15}) and courtesy of Kollipost (ν_{24}) are flanked by columns of harmonic MP2/6-31+G* predictions building on a given out-of-plane (o.o.p.) or in-plane (i.p.) bending, stretching or twisting fundamental. See also Tabs. 3.7 and 3.10.

Because of the band center uncertainties associated with thermal excitation, we have followed a slightly different prediction strategy, combining our jet-cooled Raman fundamentals with harmonic MP2/6-31+G* predictions for the IR fundamentals. The predicted combination band positions neglecting mixed anharmonicity are listed in Tab. 3.10. Some experimental data for $(HCOOD)_2$ are included as well.

We exemplify the situation for the band observed at $307/311\,\text{cm}^{-1}$. Based on the addition of IR and Raman gas phase fundamental values, two combination band assignments were proposed, namely as $\nu_{15} + \nu_9$ or as $\nu_{16} + \nu_{12}$ [9]. Both combinations give a wavenumber sum close to $300\,\text{cm}^{-1}$ $(163 + 137, 68 + 230)$ when constructed from the thermal gas phase data. By combining our accurate Raman jet band positions with MP2/6-31+G* IR predictions (or in this case equivalently the best available IR experimental data) instead, the wavenumber sum is $330\,\text{cm}^{-1}$ $(169 + 161)$ or $311\,\text{cm}^{-1}$ $(69 + 242)$. It is tempting to favor the latter assignment in this case, *i.e.* $\nu_{16} + \nu_{12}$, although the gas phase combination band maximum may be thermally shifted from the band center. However, a thermal shift of $\approx 20\,\text{cm}^{-1}$ (or more, if anharmonicity effects are taken into account) appears less likely. The predicted isotopic substitution pattern follows the experimental one quite well in both cases and cannot be used as an assignment aid in this case. On the other hand, the harmonic simulated band position of $\nu_{15} + \nu_9$ is in good agreement with the observed band at $329\,\text{cm}^{-1}$, which was not assigned in Ref. 9. Unfortunately the isotopomer bands are not observed, leaving some room for confirmation.

If we settle for the assignment of the $311\,\text{cm}^{-1}$ band to $\nu_{16} + \nu_{12}$ and of the $329\,\text{cm}^{-1}$ band to $\nu_{15} + \nu_9$, the mixed anharmonicity appears to be negligible in these cases as well. A firm statement has to await a jet-cooled IR measurement of these two combination bands. However, there is already some evidence from an analysis of hot band structure in ν_{15} caused by the lowest frequency vibration [11]. This is consistent with an anharmonicity constant $x_{15,16}$ on the order of $-1\,\text{cm}^{-1}$. The corresponding combination band $\nu_{15} + \nu_{16}$ is Raman active, but obscured by ν_{12}, as discussed above.

The relatively strong IR-active combination band at $395 \, \text{cm}^{-1}$ was previously [9] assigned as $\nu_{15} + \nu_{12}$. Based on the new data set, another possibility would be $\nu_{24} + \nu_9$, but the first assignment appears slightly more probable due to the relatively large anharmonic constant of the latter one $(-30 \, \text{cm}^{-1}$, see Tab. 3.9). IR data for either $(\text{DCOOH})_2$ or $(\text{DCOOD})_2$ can assist a firmer assignment because the wavenumber of ν_{15} is very sensitive to $\text{C}-\text{H}$ deuteration. For an assessment of anharmonicity, IR jet spectra will be indispensable.

If we assume $\nu_{15} + \nu_{12}$ as the correct assignment of the IR active band at $395 \, \text{cm}^{-1}$, it is interesting to see that only the two out-of-plane IR active fundamental bands are available for strong combinations and only the overtone bands of these two are detected with substantial intensity in our Raman jet study. The Raman out-of-plane bending mode ν_{12} is the only one of the three Raman active fundamental modes whose overtone is observed with significant intensity. One may speculate that out-of-plane motion also couples strongly to other modes and that it should be preferentially considered for the assignment of combination bands in higher frequency regions.

The most important finding is that all intermolecular modes of FAD for which combinations and overtones were observed have very small diagonal and off-diagonal anharmonicity constants, on the order of $-1 \, \text{cm}^{-1} \pm 2 \, \text{cm}^{-1}$. This is in line with reduced-dimensionality anharmonic calculations [90, 91] and appears to convey a surprisingly harmonic picture of the hydrogen bond modes in this hydrogen bond prototype. As we will show in the next section, this impression can be deceiving.

3.2.1.4 Anharmonic perturbation theory

In the recent Raman matrix isolation study [69], the standard second order perturbation theory approach implemented in the Gaussian program suite [115] was applied to ν_{12} and yielded an anharmonic wavenumber of $245 \, \text{cm}^{-1}$ at the B3LYP/6-311+G(2d,2p) level in satisfactory agreement with the experimental jet value of $242 \, \text{cm}^{-1}$. The corresponding harmonic wavenumber of $261 \, \text{cm}^{-1}$ for this mode [69] is in much poorer agreement with experiment. This can be rationalized when considering the relationship (in the absence of anharmonic resonances):

$$\nu_{12} = \omega_{12} + 2x_{12,12} + \frac{1}{2} \sum_{j \neq 12} x_{12,j} \tag{3.7}$$

The diagonal anharmonicity correction is only $-2\,\mathrm{cm}^{-1}$ whereas the five off-diagonal contributions from the other hydrogen bond modes accumulate to $-7\,\mathrm{cm}^{-1}$. Three of these are largely confirmed by experiment in the present work, the others would require far infrared jet spectroscopy. The remaining gap between the harmonic and anharmonic ν_{12} fundamental of about $-10\,\mathrm{cm}^{-1}$ is the net effect of positive and negative intra-intermolecular couplings. This underscores the need for an extensive experimental study of combination bands in the higher frequency range, because the validity of the perturbation approach also requires testing.

Tab. 3.9 summarizes the anharmonic predictions of the B3LYP/6-311+G(2d,2p) perturbational analysis and compares the effective anharmonicity constants $x_{i,j}$ with those derived from experiment. In the case of $x_{8,8}$, the anharmonicity constant of the dimer stretch, reaction surface variational results (at the B3LYP/6-31+G* level, where the fundamental may be somewhat too high in energy) are also available [90]. They are sensitive to the variational basis set, but the largest basis set yields a value of $-1.1\,\mathrm{cm}^{-1}$ in excellent agreement with experiment and reasonable agreement with perturbation theory.

Not surprisingly, an anharmonic perturbation theory analysis at the MP2/6-31+G* level reveals that the close agreement of the harmonic predictions with experiment is a consequence of fortuitous error cancellation. The (approximate) anharmonic predictions for the hydrogen bond fundamentals are now significantly lower than our experimental data.

Fig. 3.18 plots the two anharmonic predictions and the harmonic MP2/6-31+G* prediction against the experimental band centers derived in this work. The most consistent performance is found for the anharmonic B3LYP approach (circles). The harmonic MP2/6-31+G* results (squares), which are shifted by $50\,\mathrm{cm}^{-1}$ for clarity, show a perfect correlation for all but the stretching fundamental and only

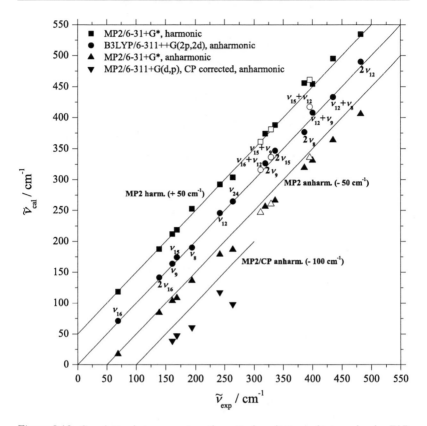

Figure 3.18: Correlation between various theoretical predictions of intermolecular FAD excitations $\tilde{\nu}_{\text{cal}}$ and experimental band centers $\tilde{\nu}_{\text{exp}}$. Filled symbols correspond to reliable band centers obtained from high resolution or jet experiments, empty symbols are more uncertain room temperature gas phase values. The squares (triangles) have been shifted up (down) by $50\,\text{cm}^{-1}$ ($100\,\text{cm}^{-1}$) for clarity.

slight overestimates for the two-quantum transitions. When corrected for anharmonicity effects, all band positions are seen to be systematically underestimated at the MP2/6-31+G* level, as expected (triangles).

In this context we wish to point out a recent quantum chemical study including anharmonicity at the MP2/6-311+G(d,p) level [43]. Comparison to the present experimental data shows that a full counterpoise (CP) correction during optimization and force field evaluation is needed to obtain qualitatively reasonable results for anharmonic constants. For example, $\nu_{12} - \omega_{12}$ is predicted at $+1.4\,\text{cm}^{-1}$ without counterpoise correction. With correction, it is predicted at $-17\,\text{cm}^{-1}$ in good agreement with the less basis set-sensitive B3LYP result of $-16\,\text{cm}^{-1}$ (6-311++G(2d,2p)). On the other hand, the absolute predictions at CP-corrected MP2/6-311+G(d,p) level are far too soft, as the corresponding trace in Fig. 3.18 shows (inverted triangles). This underlines the importance of the present benchmarks for the development of a reliable quantum chemical treatment. Limitations of the perturbation theory approach can only be assessed once a fully satisfactory electronic structure approach is employed.

3.2.1.5 The lowest intramolecular modes

The intermolecular transitions of FAD are framed on their lower end by rotational transitions, which are difficult to access in our instrument because of the Rayleigh edge filter, and on the high wavenumber end by the lowest intramolecular mode of the HCOOH dimer. This is the A_g-symmetric OCO bending mode ν_7, which has already been discussed in detail in Ref. 3. It is blue-shifted relative to the corresponding vibrations in formic acid monomer and its isotopomers (also denoted ν_7), which may be seen as weak sharp bands in the spectra of the dimer-dominated expansion (marked M in Fig. 3.12 [9,33,67]). They are easily identified by comparison to gas phase spectra (see Fig. 3.2, 3.9, 3.8 and 3.10) and by their different concentration dependence. Furthermore, they match quite closely the harmonic MP2/aug-cc-pVTZ predictions (see Tab. 3.11), whereas the fortuitously good performance of the MP2/6-31+G* approach breaks down for this and other intramolecular modes. The band in the HCOOH expansion is particularly weak because the spectrum was recorded at a stagnation pressure of 1 bar instead of 0.5 bar for the deuterated isotopomers. Therefore, the clustering extent is higher.

| Mode | Dimer | Label | Calculation (MP2) | | | Experiment |
			aug-cc-pVTZ[a]	6-31+G*	6-311+G*	
ν_7	(HCOOH)$_2$	...	684.8	670.5	679.0	682
	(DCOOH)$_2$	...	678.8	664.7	673.1	676
	(HCOOD)$_2$	...	635.0	617.5	623.6	628
	(DCOOD)$_2$	...	630.3	613.1	619.2	624
	DCOOH-DCOOD	A	...	674.2 (9.5)	679.4 (6.3)	692 (16)
	DCOOH-DCOOD	B	...	622.4 (9.3)	626.2 (2.6)	637 (13)
	HCOOH-HCOOD	C	...	680.3 (9.8)	685.5 (6.5)	698 (16)
	HCOOH-HCOOD	D	...	627.0 (9.5)	630.8 (7.2)	641 (13)
ν_{11}	(HCOOH)$_2$	...	979.4	934.3	852.8	(911[b])
	(DCOOH)$_2$	...	990.7	946.3	905.8	...
	(HCOOD)$_2$	E	721.8	693.5	639.5	672
	(DCOOD)$_2$	F	719.5	690.1	633.5	669

[a] Zielke, Ref. 33.
[b] See the next section in detail.

Table 3.11: Assignment of some intramolecular Raman active modes of (HCOOH)$_2$ and its isotopomers between 600 and 750 cm^{-1}. Experimental band maxima (in cm^{-1}) are compared with the calculated harmonic band positions (in cm^{-1}) using the MP2 method with different basis sets. The blue shifts (in cm^{-1}) from ν_7 of isotopically mixed dimers relative to the corresponding symmetric dimer modes are listed in parentheses.

Tab. 3.11 lists some weak dimer bands near the lowest intramolecular FAD vibration ν_7, which are marked with different labels in Fig. 3.12. It is unlikely that they arise from water impurities in the formic acid samples, although such impurities are always an issue in this chemically unstable compound. The bands at 692 cm^{-1} (Label: A) and 637 cm^{-1} (Label: B) in the jet spectrum of DCOOH are instead assigned as the intramolecular OCO bending modes of the unsymmetrically isotope-substituted DCOOH–DCOOD dimer. Such mixed dimers can arise from partial isotope exchange at the container walls and the presence of about 10% DCOOD is indeed also evidenced by a small DCOOD monomer peak (marked in Fig. 3.12). Weaker evidence of isotope exchange is also seen in the DCOOD and HCOOD spectra, but not in the HCOOH spectra, which were measured before introducing deuterated compounds into the apparatus. The lower frequency band at 637 cm^{-1} in the DCOOH spectrum is unusually strong, indicating that there might be another contribution underneath, e.g. from $3\nu_{12}$.

Both mixed dimer bands (A, B) are shifted by about $15\,\mathrm{cm}^{-1}$ to higher wavenumber relative to their symmetric dimer ν_7 counterparts (see Tab. 3.11). This blue-shift is underestimated by the calculations, but the assignment is still straightforward. It is supported by the fact that the A band at $692\,\mathrm{cm}^{-1}$ was also detected in the spectrum of DCOOH. Furthermore, the related symmetry-broken OCO bending bands of the HCOOH–HCOOD dimer are observed at $698\,\mathrm{cm}^{-1}$ (Label: C) and $641\,\mathrm{cm}^{-1}$ (Label: D) in the spectrum of HCOOD with very similar band shifts.

The shoulder on the high-frequency wing of the HCOOH spectrum (Fig. 3.12) may be due to ν_{23}, the IR counterpart of ν_7, in reduced symmetry clusters of dimers. An assignment to the second overtone of ν_{12} is less likely.

In O-deuterated variants of formic acid monomer, ν_7 is actually not the lowest fundamental. Instead, the strongly hindered internal rotation or torsion of the OH group relative to the molecular frame is lower, although not as low as initially believed [116]. Like the true rotations of the monomer, this torsion experiences a significant blue-shift in the hydrogen-bonded dimer. Indeed, it would merge with an intermolecular fundamental of the dimer if the mass of the formyl radical were negligible [97]. It certainly plays a major role in the isotope anomaly of the dimer stretching vibration [3]. However, this symmetric torsion has not been assigned before in deuterated FAD because of its very low Raman activity.

We would like to argue that the two remaining unassigned bands in Fig. 3.12, band E at $672\,\mathrm{cm}^{-1}$ in the spectrum of HCOOD and band F at $669\,\mathrm{cm}^{-1}$ in the spectrum of DCOOD may be due to this rather elusive out-of-plane O–D bending vibration $\nu_{11}\,(\mathrm{B}_g)$ of the dimer. The harmonic predictions for this mode show a wide variation (see Tab. 3.11) but the C–H/C–D isotope effect is relatively stable and in good agreement with the observation. Interestingly, the MP2/6-31+G* prediction is again the closest, in particular if one takes into account some negative anharmonicity contribution $x_{11,11}$ which will almost certainly exist for this large amplitude vibration. A doublet band characteristic for B_g symmetry in the non-deuterated HCOOH dimer spectrum at $911\,\mathrm{cm}^{-1}$ (see Fig. 3.2 and Tab. 3.5) is also consistent with a ν_{11} assignment. We note that it has recently been assigned in the room temperature gas phase at $922.0 \pm 1.5\,\mathrm{cm}^{-1}$ [63], confirming an earlier assignment. A thermal shift of $11\,\mathrm{cm}^{-1}$ is not unusual for such an intramolecular vibration, but the direction of the shift is surprising (see Fig. 3.2). Normally, one would expect a shift towards the corresponding monomer fundamental, which is found at much lower wavenumber. Even in the case of the less hydrogen bond affected ν_7 vibration, the thermal gas phase band maximum is significantly redshifted. Therefore, one should not dismiss other explanations for the gas phase band such as $\nu_7 + \nu_{12}$ at this stage.

A future measurement series under depolarized conditions, spontaneous jet, post-Mach disk and gas phase spectra in the $800\text{-}1000\,\mathrm{cm}^{-1}$ window may reveal the true nature of the various bands, in particular if water traces are added to identify formic acid-water complexes and if isotope substitution is included.

3.2.1.6 Conclusions

The small anharmonicity effects in the van der Waals modes tend to support reaction path Hamiltonian approaches which treat bath modes harmonically [92–94]. However, the cumulative effect of small anharmonicity contributions on the fundamental frequencies can be rather large, as anharmonic perturbation calculations indicate. Basically, all spectral features down to the noise level can be attributed to symmetric dimers and monomers of formic acid, supporting the expectation that isomeric forms of the dimer are not present in significant amounts.

Taken together, these new experimental results advance our understanding of the low frequency FAD dynamics quite substantially, as summarized in Fig. 3.17. They will also promote the assignment of the extremely complex spectra [91,108] above $2000\,\mathrm{cm}^{-1}$. An important intermediate step will be a jet study of the intramolecular Raman spectra below $1800\,\mathrm{cm}^{-1}$, which will be discussed in the next section. Even more so than in the monomers [76], this region will involve a number of vibrational resonances in the dimers, which need to be unravelled with the help of isotope substitution [95,98].

A largely improved experimental setup will be needed to resolve FAD tunneling splittings as a function of vibrational excitation, because these splittings are exceedingly small [79, 93]. However, the symmetric modes which Raman spectroscopy is able to probe are particularly promising in this context, because they are expected to show the largest enhancements [94]. Tunneling will be further slowed down in homologs of FAD like the dimer of acetic acid [17, 101, 117], where methyl group rotation must be synchronized to hydrogen transfer. The study of its low frequency modes may provide additional insights into the out-of-plane dynamics of the carboxylic acid dimer subunit. This effect will be discussed in the following two chapters.

A further exploration of unsymmetric dimers, either by isotope exchange as in this work or by chemical substitution [99, 113, 118], is indicated as soon as the symmetric dimers are fully understood, because this can provide alternative insights into the hydrogen bond-mediated coupling strengths. Furthermore, the transition from concerted to stepwise double proton transfer can be investigated in such systems [119]. Ultimately, a detailed experimental anharmonic force field [88] of these prototype systems for double hydrogen bonding should be reachable by a combination of IR and Raman spectroscopy in supersonic jets. Such force fields appear to be a prerequisite for accurate tunneling splitting predictions [90, 92, 93].

3.2.2 750-1800 cm^{-1} region

All the intramolecular fundamental modes of FAD except the O–H/D and the C–H/D stretching vibrations are below $1800 \, \text{cm}^{-1}$. The Raman active fundamentals of all the four isotopomers have been assigned in Tab. 3.5. IR active fundamentals in this region, which can also give rise to the overtone/combination bands, have been characterized a long time ago in the GP [18, 75], and investigated in jet [95]/matrix [13–15] experiments over the last decade. These previous experimental results are summarized in Tab. 3.12. ν_{20} and ν_{21} of $(\text{HCOOD})_2$ were observed in an early low resolution reference [75], but re-assigned in later work [83].

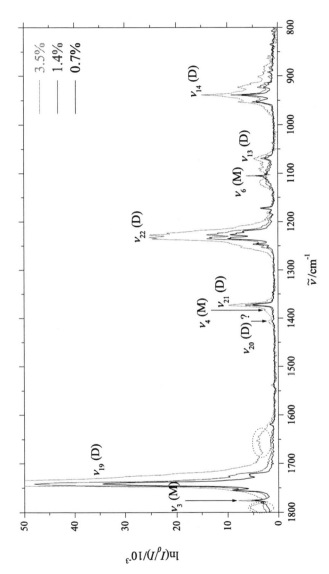

Figure 3.19: FTIR jet spectra of $(HCOOH)_2$ in He between 800 and 1800 cm^{-1} at variable conditions, measured by F. Kollipost, personal communication. $T_s = 15/0/-10°C$ (from top to bottom), $p_0 = 1.5$ bar, $p_s = 700$ mbar. The broad bands in the circles in the spectrum with high concentration are assumed to come from larger clusters.

Several fundamentals of $(HCOOH)_2$ in this wavenumber region are shown in the FTIR jet spectra in Fig. 3.19[1]. The experimental results are listed in Tab. 3.12 for comparison. The C–O stretching band ν_{22} shows a complicated band structure. The two sub-bands have nearly the same intensities and both band positions are listed in Tab. 3.12. It is seen that the band positions measured with low temperature techniques are generally shifted to high wavenumber compared to the room temperature spectra, except for the carboxyl vibration ν_{19}. Its band position observed in the Ar-matrix experiment [13, 15] is even significantly lower than that in our jet spectra, in part due to the much lower vibrational temperature in the matrix. Beside this, there is a good agreement between the jet and matrix experimental results. The wavenumber differences are largely within $4\,cm^{-1}$. A weak band at $1406\,cm^{-1}$ in our jet spectra is assigned as the IR active O–H out-of-plane bending fundamental mode ν_{20}, much lower than the band positions at around $1450\,cm^{-1}$ observed in the GP spectra at room temperature [11, 75]. Our assignment is strongly supported by the quantum chemical calculations, which will be discussed in the following text.

The experimental results of the fundamental modes of $(HCOOH)_2$ below $1800\,cm^{-1}$ are compared with the quantum-chemical calculations at several different levels using various methods (see Tab. 3.13). The best-fit method/level will be used further to assist the assignment of the combination/overtone bands. Experimental values of all the Raman active fundamentals and most IR active ones used for the comparison are from the jet measurements. The band position of the IR active C–O stretching fundamental ν_{22} was accurately determined to $1230.2\,cm^{-1}$ using the jet-cooled cavity ring-down IR spectra [95] and therefore included in the comparison in Tab. 3.13, instead of the jet value of $1227/1232\,cm^{-1}$. The IR active O–H out-of-plane bending ν_{14} is below the wavenumber region of our jet measurements and therefore its Ar matrix isolation experimental result [13] is used to compare with the calculations. The experimental values of the intermolecular fundamentals are the same as those in Tab. 3.7. The chosen experimental

---

[1]Measurements by F. Kollipost, personal communication.

Description	Mode	(HCOOH)$_2$			(DCOOH)$_2$		(HCOOD)$_2$	(DCOOD)$_2$
		GP	Jet	Ar-Matrix	GP[a]	Ar-Matrix[b]	GP	GP
	B$_u$							
ν(C=O)	ν_{19}	1754[a], 1746[c], 1740[d]	1741[e]	1729/1730[b], 1728[f]	1726	1708/1709/1723	1745[a], 1736[d]	1720[a], 1720[d], 1717.5[g]
δ(O–H)	ν_{20}	1450[a], 1454[c]	1406[e]	...	1360	1381/1384	1037[a], 1382[d]	1055[a], 1070[d]
δ(C–H)	ν_{21}	1365[a], 1364[c], 1364[d]	1373[e]	1372/1373[b], 1373[f]	996	1003/1006	1387[a]	976/987[a], 984[d]
ν(C–O)	ν_{22}	1218[a], 1218[c], 1215[d]	1227/1222[e], 1230.2[h]	1223/1227[b], 1226[f]	1239	1244	1259[a], 1249[d]	1246[a], 1249[d]
δ(OCO)	ν_{23}	697[a], 698[c], 699[d]	...	712[e]	695	...	651[a], 675[d]	642[a]
	A$_u$							
γ(C–H)	ν_{13}	1050[a]	1069[e]	1068/1072[b]	890	956/965	1037[a]	890[a]
γ(O–H)	ν_{14}	917[a], 922[c], 908[d]	939[e]	940/947[b], 942[f]	930	891/892	693[a]	678[a], 669[d]

[a] Millikan and Pitzer, IR GP spectra at room temperature (Ref. 75).
[b] Ito, Ar matrix isolation IR spectrum (Ref. 15).
[c] Georges et al., high resolution IR GP spectra (Ref. 11).
[d] Maréchal, IR GP spectra at room temperature (Ref. 83).
[e] IR jet spectrum, courtesy of Kollipost.
[f] Halupka and Sander, Ar matrix isolation IR spectrum (Ref. 13).
[g] Gutberlet et al., high resolution FIR GP spectra (Ref. 85).
[h] Ito, jet-cooled cavity ring-down IR spectra (Ref. 95).

Table 3.12: Band positions (in cm^{-1}) and assignment of the IR active fundamental modes of FAD and its isotopomers.

		B3LYP		MP2				
Mode	Exp.[b]	6-31+G*	6-311++G(2d,2p)	6-31+G*	6-311+G*	6-311+G(2d,p)	6-311+G(3d,p)	aug-cc-pVTZ[a]
Raman active[b]								
ν3	1668	1721.6	1694.1 (1648.6)	1734.7 (1695.8)	1744.0 (1708.5)	1696.5 (**1656.8**)	1705.1 (**1662.6**)	1703.2
ν4	1431	1458.8	1489.3 (1443.2)	1463.7 (**1419.8**)	1459.3 (1409.4)	1487.4 (**1442.7**)	1484.5 (**1424.6**)	1483.2
ν5	1376	1405.1	1405.8 (**1372.3**)	1411.3 (**1371.8**)	1405.3 (1362.8)	1421.3 (**1385.6**)	1410.7 (**1374.3**)	1408.2
ν6	1224	1253.2	1253.1 (**1224.8**)	1244.6 (**1212.6**)	1246.5 (**1212.1**)	1249.9 (**1218.7**)	1250.8 (**1219.8**)	1261.1
ν7	682	**676.9**	**688.7** (**679.1**)	**670.5** (663.4)	**679.0** (**670.4**)	**688.0** (**680.4**)	**682.4** (**675.6**)	**684.8**
ν8	194	209.3	207.1 (**190.2**)	**202.9** (**186.5**)	**191.6** (175.8)	**199.6** (**185.5**)	**202.9** (**188.2**)	212.3
ν9	161	**169.5**	173.7 (**163.9**)	**162.1** (**153.5**)	**159.1** (148.2)	**165.5** (**153.4**)	**168.8** (**157.7**)	**169.1**
ν10	1060	**1068.8**	1077.3 (**1050.9**)	1078.5 (**1053.1**)	1079.3 (**1060.9**)	1082.2 (**1055.7**)	1085.2 (**1048.3**)	1089.4
ν11	911	953.3	989.0 (931.8)	934.3 (869.3)	852.8 (888.8)	975.5 (**911.3**)	988.2 (888.1)	979.4
ν12	242	255.1	261.8 (**246.0**)	**242.3** (228.8)	209.0 (**236.2**)	257.5 (**244.2**)	259.5 (**240.8**)	260.0
IR active								
ν13	1069[c]	1089.1	1103.0 (**1060.2**)	1098.6 (**1061.5**)	1091.3 (**1069.8**)	1107.6 (**1068.2**)	1113.7 (1050.7)	1116.4
ν14	939[c]	981.4	1008.6 (967.3)	964.8 (912.1)	882.1 (**945.0**)	995.7 (**950.3**)	1005.6 (**933.3**)	997.8
ν15	168.5[d]	**178.5**	186.8 (**174.4**)	**169.0** (**158.5**)	142.9 (**162.8**)	**176.3** (**169.0**)	182.8 (**162.3**)	183.3
ν16	69.2[d]	**76.0**	**77.8** (**71.3**)	**68.8** (**67.4**)	**63.7** (**70.6**)	**70.1** (**67.2**)	**69.1** (**66.5**)	**69.3**
ν19	1741[c]	1785.4	1765.5 (1725.2)	1782.9 (**1746.9**)	1790.7 (1756.5)	1761.0 (1723.6)	1767.7 (**1729.2**)	1770.9
ν20	1406[c]	1440.3	1457.2 (**1407.3**)	1454.5 (**1413.7**)	1449.6 (**1408.4**)	1462.8 (**1413.6**)	1459.2 (**1406.1**)	1455.7
ν21	1373[c]	1396.6	1404.4 (**1371.4**)	1398.2 (1355.7)	1393.3 (1348.6)	1414.3 (**1375.2**)	1404.1 (**1366.9**)	1404.1
ν22	1230.2[e]	1254.6	1257.7 (**1233.3**)	1247.4 (1217.9)	1250.2 (1215.8)	1255.9 (**1229.8**)	1257.5 (**1226.9**)	1265.7
ν23	712[f]	**702.9**	**722.9** (**714.6**)	686.6 (677.0)	690.2 (680.4)	**717.7** (**706.8**)	**713.7** (**702.7**)	**715.2**
ν24	264[c]	**268.3**	**274.3** (**264.6**)	**253.7** (236.9)	237.6 (219.6)	**267.2** (**254.1**)	**267.4** (**255.3**)	281.2

[a] Wavenumbers from Zielke, Ref. 33.
[b] Raman jet spectrum, this work.
[c] IR jet spectrum, courtesy of Kollipost, Ref. 33.
[d] Georges et al., high resolution FIR GP spectrum (Ref. 11).
[e] Ito, jet-cooled cavity ring-down IR spectra (Ref. 95).
[f] Halupka and Sander, Ar matrix isolation IR spectrum (Ref. 13).

Table 3.13: Comparison of calculated (HCOOH)$_2$ fundamentals below 1800 cm^{-1} using different quantum chemical methods and basis sets with experimental values (in cm^{-1}). The anharmonic calculation results are listed in parentheses, whereas the calculations with wavenumber shifts within 12 cm^{-1} of the related experimental values are given in bold face.

wavenumbers are also used for all the later predictions of the combination/overtone bands.

For the intermolecular modes, the harmonic calculation at the simple MP2/6-31+G* level fits best to the experimental results, even compared to the anharmonic calculations(see Tab. 3.13). It is not surprising that it does not fit well to the intramolecular modes in this wavenumber range because it is already about 2% underestimated for the lowest intramolecular mode ν_7. Actually no harmonic calculation shows a satisfactory agreement. For the vibrations above $900\,\mathrm{cm}^{-1}$, both IR and Raman active, the harmonic calculated wavenumbers are generally more than $20\,\mathrm{cm}^{-1}$ overestimated, even at the high B3LYP/6-311++G(2d,2p), MP2/6-311+G(3d,p) and MP2/aug-cc-pVTZ levels (see Tab. 3.13). In contrast, the anharmonic calculations show a much better agreement: most calculated wavenumbers at the three highest levels (B3LYP/6-311++G(2d,2p), MP2/6-311+G(2d,p) and MP2/6-311+G(3d,p)) shift less than $12\,\mathrm{cm}^{-1}$ from the experimental results, which means a deviation on the order of 1% or less. The only exception is the IR active carboxyl stretching mode ν_{19}. The anharmonic calculations fit better with the Ar matrix experiments around $1730\,\mathrm{cm}^{-1}$ [13, 15]. The band position of the IR active O–H out-of-plane bending mode ν_{20} was reported at about $1450\ \mathrm{cm}^{-1}$ in the GP spectra at room temperature [11,75], which fits all the harmonic calculations quite well. However, no band with a similar wavenumber to that mentioned above but only a weak band at $1406\,\mathrm{cm}^{-1}$ was observed in our jet spectra. All the anharmonic calculations show a very good agreement with this wavenumber. This band is so weak in our jet spectra because it has the smallest calculated IR intensity among all the intramolecular fundamental modes.

Assisted by calculations we are able to assign most of the numerous overtone/combination bands of $(HCOOH)_2$ in this wavenumber range. The band positions and assignments are listed in Tab. 3.14, compared with the anharmonic calculation results at the MP2/6-311+G(2d,p) level. This level was chosen for the comparison because it offers not only a good fit of the band positions for $(HCOOH)_2$ but it is the one also among the levels in Tab. 3.13 that shows the best agreement of the wavenumber shifts between $(HCOOH)_2$ and its three

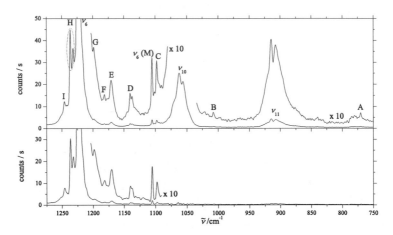

Figure 3.20: Depolarization analysis of $(HCOOH)_2$ between 750 and 1275 cm^{-1}, with the excitation laser perpendicular to the scattering plane (top trace) and residual after subtracting 7/6 of the spectrum obtained with the excitation laser parallel to the scattering plane. $T_s = 20°C$ (4.5% HCOOH in He), $p_s = 700$ mbar, $d = 0.4$ mm, 12×200 s, 4.0×0.15 mm^2 nozzle.

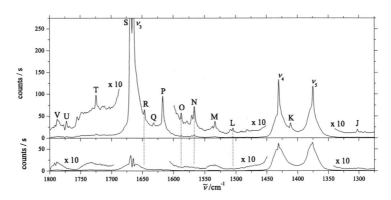

Figure 3.21: Depolarization analysis of $(HCOOH)_2$ between 1275 and 1800 cm^{-1}, with the excitation laser perpendicular to the scattering plane (top trace) and residual after subtracting 7/6 of the spectrum obtained with the excitation laser parallel to the scattering plane. $T_s = 16°C$ (3.7% HCOOH in He), $p_s = 700$ mbar, $d = 0.5$ mm, 8×200 s.

isotopomers. Therefore, the results calculated at this level will be used for the assignments of the overtone/combination bands of the other three FAD isotopomers later on.

The band symmetry is determined by depolarization measurements, shown in Fig. 3.20 and Fig. 3.21. The overtone/combination bands are numbered with labels A to V, which are also used in Tab. 3.14. Since all overtone/combination bands down to the noise level in the $\leq 750\,\mathrm{cm}^{-1}$ region (see Fig. 3.15) can be basically attributed to binary combinations, this will also be the preferred assumption for the assignment of the combination bands in the other wavenumber regions. Two conditions must be fulfilled for the proposed assignments: same symmetry and similar wavenumbers of the (harmonic) combined bands and the (anharmonic) experimental ones. Normally, there is more than one possible combination for the observed bands. In this case, if the symmetry of some bands is very clear, only the combination with the right symmetry is chosen. For some very weak bands, the band symmetry remains uncertain and all possible combinations will be listed. Sometimes there is only one possible way to get a (harmonically estimated) combination band with a similar band position as the experimental one, and then the symmetry of the band can be inferred. The band symmetries determined in this indirect way are shown in bold face in Tab. 3.14.

It should be noted that both depolarization analysis were carried out with relatively high acid concentration. For a safe exclusion of A_g symmetry of several bands, e.g., the band marked ν_{11}, a measurement at lower concentration would be helpful.

The predictions are made based on the relatively accurate experimental IR active fundamental modes (jet/high resolution GP/Ar matrix) and the jet Raman data of this work, neglecting mixed anharmonicity contributions. The wavenumber difference of the sum of the two partners, relative to the experimental values, is listed in parentheses. Ideally, it can be interpreted as the experimental "anharmonicity constant" of the combination band $(x_{i,j})$. However, it may be contaminated by errors in the IR active fundamentals. Generally, there is a good agreement among the experimental values of this work, the anharmonic calcula-

Label	Experiment	Symmetry	Assignment	Prediction	Calculation
A	770	B_g	$\nu_{16}+\nu_{23}$	$69.2^a + 712^b = 781 \ (-11)$	$773.0 \ (-1)$
B	1008	A_g	$\nu_{14}+\nu_{16}$	$939^c + 69.2^a = 1008 \ (0)$	$1016.6 \ (-1)$
C	1099	A_g	$\nu_{14}+\nu_{15}$	$939^c + 168.5^a = 1108 \ (-9)$	$1110.5 \ (-9)$
D	1137/1140	A_g	$\nu_{13}+\nu_{16}$	$1069^c + 69.2^a = 1138$	$1134.4 \ (-1)$
		A_g	$\nu_{11}+\nu_{12}$	$911^d + 242^d = 1153$	$1139.8 \ (-16)$
E	1172	A_g	?		
F	1183	A_g	?		
G	1198	A_g	?		
H	1232/1237	A_g	$\nu_{13}+\nu_{15}$	$1069^c + 168.5^a = 1238$	$1232.2 \ (-5)$
I	1246	A_g	$\nu_{13}+\nu_{15}$	$1069^c + 168.5^a = 1238 \ (8)$	$1232.2 \ (-5)$
J	1307		ν_5 (M)		1253.6
		B_g	$\nu_{16}+\nu_{22}$	$69.2^a + 1230.2^e = 1299 \ (8)$	$1290.7 \ (-6)$
		B_g	$\nu_{13}+\nu_{24}$	$1069^c + 264^c = 1333 \ (-26)$	$1315.5 \ (-7)$
K	1412	A_g	$\nu_6+\nu_8$	$1224^d + 194^d = 1418 \ (-6)$	$1400.9 \ (-3)$
L	1505	B_g	$\nu_{16}+\nu_{20}$	$69.2^a + 1406^c = 1475 \ (30)$	$1479.5 \ (-1)$
		A_g	$\nu_{22}+\nu_{24}$	$1230.2^e + 264^c = 1494 \ (11)$	$1473.4 \ (-11)$
M	1534	A_g	$\nu_5+\nu_9$	$1376^d + 161^d = 1537 \ (-3)$	$1540.0 \ (1)$
N	1567	A_g	$\nu_5+\nu_8$	$1376^d + 194^d = 1570 \ (-3)$	$1569.6 \ (-2)$
O	1588	A_g	$\nu_4+\nu_9$	$1431^d + 161^d = 1592 \ (-4)$	$1592.4 \ (-4)$
		B_g	$\nu_7+\nu_{11}$	$682^d + 911^d = 1593 \ (-5)$	$1588.5 \ (-3)$
P	1617	B_g	$\nu_5+\nu_{12}$	$1376^d + 242^d = 1618 \ (-1)$	$1629.2 \ (-1)$
		B_g	$\nu_{15}+\nu_{20}$	$168.5^a + 1406^c = 1575 \ (-6)$	$1581.0 \ (-2)$
Q	1632	B_g	$\nu_{14}+\nu_{23}$	$939^c + 712^b = 1651 \ (-19)$	$1654.7 \ (-2)$
R	1646	B_g	$\nu_{14}+\nu_{23}$	$939^c + 712^b = 1651 \ (-6)$	$1654.7 \ (-2)$
		B_g	$\nu_4+\nu_{12}$	$1431^d + 242^d = 1673 \ (-27)$	$1682.6 \ (-4)$
S	1669	A_g	$\nu_{20}+\nu_{24}$	$1406^c + 264^c = 1670 \ (-1)$	$1662.9 \ (-5)$
T	1725	B_g	$\nu_7+\nu_{10}$	$682^d + 1059^d = 1741 \ (-16)$	$1735.5 \ (-1)$
U	1773	B_g	$\nu_{13}+\nu_{23}$	$1069^c + 712^b = 1781 \ (-8)$	$1771.7 \ (-3)$
V	1788	B_g	$\nu_{16}+\nu_{19}$	$69.2^a + 1741^c = 1810 \ (-22)$	$1790.6 \ (0)$

[a] Georges et al., high resolution FIR GP spectrum (Ref. 11).
[b] Halupka and Sander, Ar matrix isolation IR spectrum (Ref. 13).
[c] IR jet spectrum, courtesy of Kollipost.
[d] Raman jet spectrum, this work.
[e] Ito, jet-cooled cavity ring-down IR spectra (Ref. 95).

Table 3.14: Band positions (in cm^{-1}) and assignment of the Raman active overtone/combination bands of $(HCOOH)_2$ between 750 and 1800 cm^{-1}, compared with the anharmonic calculation at the MP2/6-311+G(2d,p) level (in cm^{-1}). The band positions of the fundamentals are listed in Tabs. 3.5 and 3.12. The experimentally and theoretically derived "anharmonicity constants" of the combination bands ($x_{i,j}$, see Equas. 3.5 and 3.6) are given in parentheses. See text for more details.

tion and the prediction. The "nominal anharmonicity constants" are satisfactorily small for the combination of two Raman active fundamentals, largely within $6 \, \text{cm}^{-1}$ and at most $16 \, \text{cm}^{-1}$. The combination bands of two IR modes usually have larger anharmonicity constants up to $30 \, \text{cm}^{-1}$.

The wavenumber difference between the anharmonically calculated band position of the combination/overtone band and the sum of those of the two fundamentals is also given in Tab. 3.14, as the theoretically derived "anharmonicity constants". Most of them are negative and small (within $16 \, \text{cm}^{-1}$ for all the bands in Tab. 3.14). In some cases, comparisons of the "experimental" and "theoretical" anharmonicity constants supply helpful information to assist the assignment. For example, both bands H and I have the similar wavenumbers to the predicted combination band $\nu_{13} + \nu_{15}$, but the negative theoretical anharmonicity constant of this band indicates that band H fits better. Nevertheless, both possibilities are listed in Tab. 3.14 although one of them is clearly preferred, because as much as possible information would be given in this work as reference for the more accurate and detailed band assignment in the future.

Several things are notable. The three bands E, F, G clearly correspond to A_g symmetry according to the depolarization experiment but there are no suitable combination bands from the prediction/calculation with A_g symmetry and similar wavenumbers. The bands persist at high dilution (see Fig. 3.22), so an assignment to larger cluster is less likely. Only the two combination bands $\nu_9 + \nu_{10}$ and $\nu_{14} + \nu_{24}$ have a similar wavenumber, but both have B_g symmetry. These bands may come from combinations of more than two bands.

Three bands were already observed and assigned in the earlier Raman GP work [9]. The weak band J at $1307 \, \text{cm}^{-1}$ in our jet spectrum was observed at $1307.1 \, \text{cm}^{-1}$ and assigned as the overtone of ν_9 of HCOOH monomer. This assignment is less likely. Detailed discussion can be found in Chap. 3.1.1. The jet spectrum in Fig. 3.20 was carried out with a high acid concentration (4.5% HCOOH in He) and the band at $1307 \, \text{cm}^{-1}$ is so weak that it could be the strong band at $1308 \, \text{cm}^{-1}$ in the GP spectrum, which is believed to come from the HCOOH-H_2O-complex. Beside this possibility, two combination bands $(\nu_{16} + \nu_{22})/(\nu_{13} + \nu_{24})$

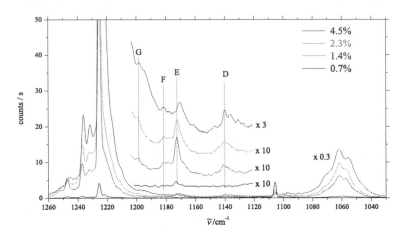

Figure 3.22: Raman jet spectra of (HCOOH)$_2$ in He between 1030 and 1260 cm^{-1} at variable conditions. From top to bottom, T_s = 20°C, p_s = 1100 mbar, d = 0.3 mm, 8 × 300 s; T_s = 8°C, p_s = 500 mbar, d = 1 mm, 8 × 300 s; T_s = 0°C, p_s = 500 mbar, d = 1 mm, 8 × 300 s; T_s = −10°C, p_s = 500 mbar, d = 1 mm, 8 × 600 s. All the spectra were carries out with the 67 L reservoir, the 250 m^3/h Roots pump and the 100 m^3/h rotary vane pump, and the 4.0 × 0.15 mm^2 nozzle. The labels are the same as those in Fig. 3.20.

with similar band positions are also listed in Tab. 3.14. The band M was observed at 1510.8 cm^{-1} in the Raman GP spectrum [9], and given the same assignment of $\nu_5 + \nu_9$ as in this work. Finally, the doublet structure of the carboxyl stretching vibration ν_3 was not distinguished but a shoulder at 1692.1 cm^{-1} was observed in the Raman GP work [9] due to the low resolution of 10 cm^{-1}, which is much larger than the wavenumber difference of 4 cm^{-1} between the two sub-bands. Fermi resonance between ν_3 and an overtone/combination band with the same A$_g$ symmetry and similar wavenumber should be responsible for the high intensity and the same band type of the two sub-bands. The shoulder was assigned as $\nu_{20} + \nu_{24}$ [9] in the Raman GP work [9], which remains valid in this work.

The Fermi-triad system in the IR active C–O stretching region ν_{22} of (HCOOH)$_2$ was analyzed recently, using jet-cooled cavity ring-down IR spectroscopy [95]. The three bands at $1220.5\,\mathrm{cm}^{-1}$, $1226.0\,\mathrm{cm}^{-1}$ and $1234.7\,\mathrm{cm}^{-1}$ were assigned as the perturbers $\nu_{12} + \nu_{14}$, $\nu_{10} + \nu_{15}$ and the fundamental ν_{22}, respectively. The unperturbed related bands were fixed to $1221.2\,\mathrm{cm}^{-1}$, $1229.6\,\mathrm{cm}^{-1}$ and $1230.2\,\mathrm{cm}^{-1}$ after analysis of the Fermi-triad coupling. The wavenumbers of the three related bands were calculated at the MP2 level for various basis sets. $\nu_{12} + \nu_{14}$ was considered as a possible resonance partner because it has a similar band position as the other two bands from some harmonic calculations, for example, at the MP2/6-311+G(2d,p) level, though the wavenumber shifts are relatively large ($> 20\,\mathrm{cm}^{-1}$) at most levels. However, at the anharmonic MP2/6-311+G(2d,p) level, which fits best to most experimental results, the wavenumber of the combination band $\nu_{12} + \nu_{14}$ is $1186.6\,\mathrm{cm}^{-1}$, more than $30\,\mathrm{cm}^{-1}$ shifted relative to the other two bands calculated at the same level: $1217.6\,\mathrm{cm}^{-1}$ for $\nu_{10} + \nu_{15}$ as well as $1229.8\,\mathrm{cm}^{-1}$ for ν_{22}. Besides the calculated profiles, the following predictions show the small likelihood of $\nu_{12} + \nu_{14}$ as a Fermi resonance partner: $242 + 939 = 1181\,\mathrm{cm}^{-1}$ for $\nu_{12} + \nu_{14}$; $1055/1062 + 168.5$ [11] $= 1223.5/1230.5\,\mathrm{cm}^{-1}$ for $\nu_{10} + \nu_{15}$. Therefore, we agree that $\nu_{10} + \nu_{15}$ and ν_{22} contribute to the Fermi resonance system analyzed in Ref. 95 but $\nu_{12} + \nu_{14}$ is less possible. A three-fold structure is also observed in the Raman active C–O stretching (ν_6) region of $(\mathrm{HCOOH})_2$. The strong fundamental band is assigned at $1224\,\mathrm{cm}^{-1}$, and the two other bands at $1232/1237\,\mathrm{cm}^{-1}$ (marked with H) are assigned as $\nu_{13} + \nu_{15}$, which is the corresponding combination band of $\nu_{10} + \nu_{15}$ and supports the analysis above. The corresponding band of $\nu_{12} + \nu_{14}$ is $\nu_{11} + \nu_{12}$ with the predicted band position of $1153\,\mathrm{cm}^{-1}$ (see Tab. 3.14), which is too far from the ν_6 and $\nu_{13} + \nu_{15}$ bands.

The overtone/combination bands of three deuterated isotopomers in this wavenumber range are more complicated to assign because some weak bands can also come from the mixed dimers of different monomer isotopomers, or even from clusters of acid and $(\mathrm{H/D})_2\mathrm{O}$, due to the 95% chemical purity of the deuterated compounds. The bending band of $\mathrm{D}_2\mathrm{O}$ at $1178\,\mathrm{cm}^{-1}$ is clearly observed in the jet spectra of DCOOD, although with weak band intensity. The lack of accurate experimental values of the IR active fundamentals increases the difficulty. In this case, the calculated wavenumbers at the anharmonic MP2/6-311+G(2d,p) level are used as an assignment aid.

Nevertheless, most of the overtone/combination bands could be tentatively assigned. The assignments and predictions are listed in Tab. 3.15 for $(DCOOH)_2$, Tab. 3.16 for $(HCOOD)_2$ and Tab. 3.17 for $(DCOOD)_2$, using the same labels as those in the related depolarization spectra: Figs. 3.23 and 3.24 for $(DCOOH)_2$, Figs. 3.25 and 3.26 for $(HCOOD)_2$, and Figs. 3.27 and 3.28 for $(DCOOD)_2$. Due to the good agreement in the band positions of the IR active fundamentals of $(HCOOH)_2$ observed in our FTIR jet spectra and in the Ar-matrix experiment [13, 15] (see Tab. 3.12), several IR active fundamentals of $(DCOOH)_2$ observed in the Ar-matrix experiment [15] are used for the band predictions in the following assignments. The anharmonic fundamental band positions calculated at the MP2/6-311+G(2d,p) level are used when the relatively accurate experimental band positions are unavailable.

Few experimental results of overtone/combination bands in this region have been reported, most of which focused on the carbonyl stretch region. Broad wavenumber ranges and complicated band structures are shown for the carboxyl stretch vibration band in the jet spectra of all the three isotopomers, especially for the two O-deuterated compounds, whose C=O band structures are not much simpler than those of the C–H(D)/O–H(D) bands. Several bands around the carboxyl stretch vibration band were observed and preliminarily assigned in the previous Raman GP studies: $1662.8/1679.2\,\mathrm{cm}^{-1}$ of $(HCOOD)_2$, which were assigned as $\nu_3/\nu_{20} + \nu_{23}$ [67] and $1684\,\mathrm{cm}^{-1}$ of $(DCOOD)_2$, which was assigned as $\nu_5 + \nu_7$ [9]. Another band at $1632.4\,\mathrm{cm}^{-1}$ was also observed in the Raman GP spectrum of $(DCOOD)_2$, but not assigned. In the Raman jet spectra, more subbands are observed due to the better resolution. Most of these numerous bands cannot be assigned directly by overtone/combination bands of only two components. The perturbations should be attributed to some further available interaction mechanisms, e.g., additional Fermi resonance as well as Coriolis coupling. The last one was proposed as an important reason for the set of carboxylic stretch vibration bands of the monomer DCOOH/DCOOD [76, 77, 120].

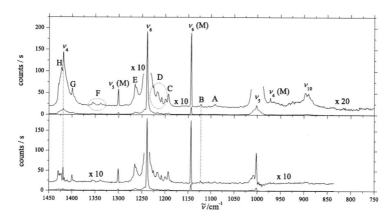

Figure 3.23: Depolarization analysis of $(DCOOH)_2$ between 750 and 1450 cm^{-1}, with the excitation laser perpendicular to the scattering plane (top trace) and residual after subtracting 7/6 of the spectrum obtained with the excitation laser parallel to the scattering plane. $T_s = 0°C$ (1.4% DCOOH in He), $p_s = 700$ mbar, $d = 0.4$ mm, 8×200 s.

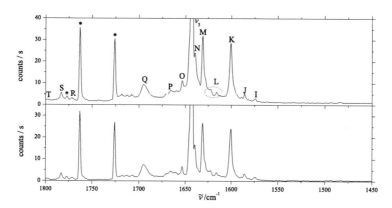

Figure 3.24: Depolarization analysis of $(DCOOH)_2$ between 1450 and 1800 cm^{-1}, with the excitation laser perpendicular to the scattering plane (top trace) and residual after subtracting 7/6 of the spectrum obtained with the excitation laser parallel to the scattering plane. $T_s = 0°C$ (1.4% DCOOH in He), $p_s = 700$ mbar, $d = 0.4$ mm, 8×150 s. The bands marked with $*$ are from DCOOH monomer but not from fundamental modes. See Fig. 3.8 for more details.

Table 3.15: Band positions (in cm^{-1}) and assignment of the Raman active overtone/combination bands of (DCOOH)$_2$ between 750 and 1800 cm^{-1}, compared with the anharmonic calculation at the MP2/6-311+G(2d,p) level (in cm^{-1}). The experimentally and theoretically derived "anharmonicity constants" of the combination bands ($x_{i,j}$) are given in parentheses. See text for more details.

Label	Experiment	Symmetry	Assignment	Prediction	Calculation
A	1091	$\mathbf{A}_g$	$\nu_{13} + \nu_{15}$	956/965[a] + 144.8[b] = 1101/1110	1096.9 (−11)
		B_g	$\nu_8 + \nu_{10}$	193[c] + 894[c] = 1087 (4)	1071.4 (−4)
		A_g	$\nu_{10} + \nu_{12}$	894[c] + 212[c] = 1106 (−15)	1100.8 (−4)
B	1122	$\mathbf{B}_g$	$\nu_8 + \nu_{11}$	193[c] + 921[c] = 1114 (−23)	1091.7 (−10)
C	1193	A_g	$\nu_{11} + \nu_{12}$	921[c] + 212[c] = 1133 (−11)	1116.5 (−15)
D	1200/1207/1217/1225	A_g	$\nu_5 + \nu_8$	1001[c] + 193[c] = 1194 (−1)	1194.2 (−1)
E	1266	A_g	$\nu_{21} + \nu_{24}$	1003/1006[a] + 248.7[b] = 1252/1255	1262.4 (−1)
		A_g	?		
F	1338/1353	A_g	$2\nu_7$	2 × 676[c] = 1352	1348.3 (0)
G	1399	A_g	$\nu_6 + \nu_9$	1238[c] + 159[c] = 1397 (2)	1379.0 (−2)
		A_g	$2\nu_{23}$	2 × 700.6[b] = 1401 (−2)	1397.8 (−3)
H	1423	A_g	$\nu_6 + \nu_8$	1238[c] + 193[c] = 1431 (−8)	1409.1 (−3)
I	1574	A_g	$\nu_4 + \nu_9$	1418[c] + 159[c] = 1577 (−3)	1586.6 (−4)
J, K	1586, 1601	A_g	$\nu_4 + \nu_8$	1418[c] + 193[c] = 1611	1607.1 (−15)
L, M	1616/1621, 1631	A_g	$\nu_{20} + \nu_{24}$	1381/1384[a] + 248.7[b] = 1630/1633	1620.8 (−8)
N, O, P	1640, 1654, 1666	A_g	$\nu_5 + \nu_7$	1001[c] + 676[c] = 1677	1685.5 (0)
Q	1695	A_g	$\nu_{21} + \nu_{23}$	1003/1006[b] + 700.6[b] = 1704/1707	1714.7 (0)
R, S	1772, 1783	A_g	$2\nu_{10}$	2 × 894[c] = 1788	1774.4 (−4)
		A_g	$2\nu_{14}$	2 × 891/892[a] = 1782/1784	1775.0 (−1)
		A_g	$\nu_3 + \nu_9$	1643[c] + 159[c] = 1802 (−4)	1779.7 (−8)
T	1798	A_g	$\nu_{10} + \nu_{11}$	894[c] + 921[c] = 1815 (−17)	1803.6 (−6)

[a] Ito, Ar matrix isolation IR spectrum (Ref. 15).
[b] Anharmonic calculation results at the MP2/6-311+G(2d,p) level.
[c] Raman jet spectrum, this work.

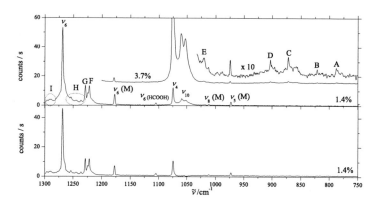

Figure 3.25: Depolarization analysis of (HCOOD)$_2$ between 750 and 1300 cm^{-1}, with the excitation laser perpendicular to the scattering plane (top trace) and residual after subtracting 7/6 of the spectrum obtained with the excitation laser parallel to the scattering plane. $T_s = 0°$C (1.4% HCOOD in He), $p_s = 700$ mbar, $d = 0.4$ mm, 8×150 s. The spectrum with 3.7% HCOOD in He in top trace is taken from Fig. 3.9.

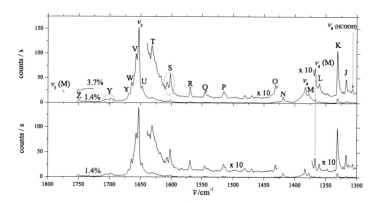

Figure 3.26: Depolarization analysis of (HCOOD)$_2$ between 1300 and 1800 cm^{-1}, with the excitation laser perpendicular to the scattering plane (top trace) and residual after subtracting 7/6 of the spectrum obtained with the excitation laser parallel to the scattering plane. $T_s = 0°$C (1.4% HCOOD in He), $p_s = 700$ mbar, $d = 0.4$ mm, 8×150 s. The spectrum with 3.7% HCOOD in He in top trace is taken from Fig. 3.9.

Label	Experiment	Symmetry	Assignment	Prediction	Calculation
A	785	A_g	$\nu_7 + \nu_9$	$628^a + 157^a = 785\ (0)$	$778.4\ (-1)$
B	821	A_g	$\nu_7 + \nu_8$	$628^a + 194^a = 822\ (-1)$	$813.3\ (-2)$
C	872	A_g	$\nu_{14} + \nu_{15}$	$723.2^b + 162.2^b = 885\ (-13)$	$876.5\ (-9)$
D	903	A_g	$\nu_{23} + \nu_{24}$	$662.8^b + 245.7^b = 909\ (-6)$	$902.5\ (-6)$
		A_g	$\nu_{11} + \nu_{12}$	$674^a + 238^a = 912\ (-9)$	$905.4\ (-12)$
E	1020	?	?		
F	1221	A_g	$\nu_{13} + \nu_{15}$	$1055.8^b + 162.2^b = 1218\ (3)$	$1217.1\ (-1)$
G	1229	A_g	$\nu_4 + \nu_9$	$1076^a + 157^a = 1233\ (-4)$	$1244.7\ (-3)$
H	1237/1244/1254	A_g	$2\nu_7$	$2 \times 628^a = 1256$	$1256.2\ (-1)$
		A_g	$\nu_4 + \nu_8$	$1076^a + 194^a = 1270$	$1279.6\ (-4)$
I	1289/1292/1296	A_g	$\nu_{10} + \nu_{12}$	$1057^a + 238^a = 1295$	$1293.7\ (0)$
		A_g	$\nu_{20} + \nu_{24}$	$1056.4^b + 245.7^b = 1302$	$1297.8\ (-4)$
J	1317	A_g	$2\nu_{23}$	$2 \times 662.8^b = 1326\ (-4)$	$1320.6\ (-3)$
K	1331	A_g	$2\nu_{11}$	$2 \times 674^a = 1348\ (-9)$	$1326.8\ (-13)$
L	1360	A_g	$2\nu_{11}$	$2 \times 674^a = 1348\ (6)$	$1326.8\ (-13)$
M	1377	A_g	?		
N	1419	A_g	$\nu_6 + \nu_9$	$1268^a + 157^a = 1425\ (-6)$	$1412.1\ (-2)$
O	1432	A_g	$2\nu_{14}$	$2 \times 723.2^b = 1446\ (-7)$	$1420.8\ (-13)$
P	1516	A_g	$\nu_{22} + \nu_{24}$	$1260.9^b + 245.7^b = 1506\ (10)$	$1502.9\ (-4)$
Q	1545	A_g	$\nu_5 + \nu_9$	$1382^a + 157^a = 1539\ (6)$	$1544.2\ (0)$
R	1569	A_g	$\nu_5 + \nu_8$	$1382^a + 194^a = 1576\ (-7)$	$1579.8\ (0)$
S	1602/1607	A_g	?		
T	1632	A_g	$\nu_{21} + \nu_{24}$	$1399.9^b + 245.7^b = 1646\ (-14)$	$1645.4\ (0)$
U	1648	A_g	$\nu_{21} + \nu_{24}$	$1399.9^b + 245.7^b = 1646\ (2)$	$1645.4\ (0)$
V	1658	A_g	?		
W	1665	A_g	?		
X	1669	A_g	?		
Y	1698/1704/1707	A_g	$\nu_4 + \nu_7$	$1076^a + 628^a = 1704$	$1721.5\ (-4)$
		A_g	$\nu_{20} + \nu_{23}$	$1056.4^b + 662.8^b = 1719$	$1714.3\ (-5)$
		A_g	$\nu_{10} + \nu_{11}$	$1057^a + 674^a = 1731$	$1721.5\ (-7)$
Z	1750	A_g	$\nu_{13} + \nu_{14}$	$1055.8^b + 723.2^b = 1779\ (-29)$	$1777.8\ (-1)$

^{a} Raman jet spectrum, this work.
^{b} Anharmonic calculation results at the MP2/6-311+G(2d,p) level.

Table 3.16: Band positions (in cm^{-1}) and assignment of the Raman active overtone/combination bands of $(HCOOD)_2$ between 750 and 1800 cm^{-1}, compared with the anharmonic calculation at the MP2/6-311+G(2d,p) level (in cm^{-1}). The experimentally and theoretically derived "anharmonicity constants" of the combination bands $(x_{i,j})$ are given in parentheses. See text for more details.

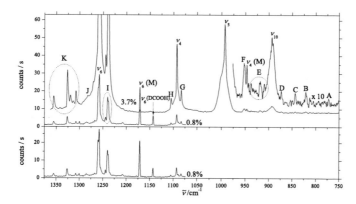

Figure 3.27: Depolarization analysis of $(DCOOD)_2$ between 750 and 1375 cm^{-1}, with the excitation laser perpendicular to the scattering plane (top trace) and residual after subtracting 7/6 of the spectrum obtained with the excitation laser parallel to the scattering plane. $T_s = -8°C$ (0.8% DCOOD in He), $p_s = 700$ mbar, $d = 0.4$ mm, 8×150 s. The spectrum with 3.7% DCOOD in He in top trace is taken from Fig. 3.10.

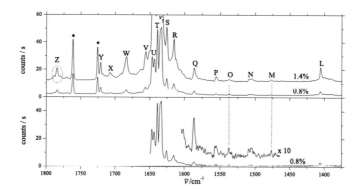

Figure 3.28: Depolarization analysis of $(DCOOD)_2$ between 1375 and 1800 cm^{-1}, with the excitation laser perpendicular to the scattering plane (top trace) and residual after subtracting 7/6 of the spectrum obtained with the excitation laser parallel to the scattering plane. $T_s = -8°C$ (0.8% DCOOD in He), $p_s = 700$ mbar, $d = 0.4$ mm, 8×150 s. The spectrum with 1.4% DCOOD in He in top trace is taken from Fig. 3.10. The bands marked with * are from DCOOD monomer but not from fundamental modes. See Fig. 3.10 for more details.

Label	Experiment	Symmetry	Assignment	Prediction	Calculation
A	773	A_g	$\nu_7 + \nu_9$	$624^a + 157^a = 781$ (-8)	773.4 (-1)
		A_g	$\nu_{14} + \nu_{16}$	$717.9^b + 67.9^b = 785$ (-12)	784.4 (-1)
		B_g	$\nu_{15} + \nu_{23}$	$140.3^b + 658.0^b = 798$ (-25)	796.5 (-2)
B	821	B_g	$\nu_9 + \nu_{11}$	$157^a + 668^a = 825$ (-4)	820.0 (-4)
		B_g	$\nu_7 + \nu_{12}$	$624^a + 210^a = 834$ (-13)	833.9 (-1)
C	843	A_g	$\nu_{14} + \nu_{15}$	$717.9^b + 140.3^b = 858$ (-15)	850.2 (-8)
		B_g	$\nu_8 + \nu_{11}$	$192^a + 668^a = 860$ (-17)	851.8 (-7)
D	874	A_g	$\nu_{11} + \nu_{12}$	$668^a + 210^a = 878$ (-4)	874.4 (-11)
		A_g	$\nu_{23} + \nu_{24}$	$658.0^b + 240.4^b = 898$ (-24)	892.5 (-6)
E	910/919/938	?	?		
F	952	A_g	$\nu_{13} + \nu_{16}$	$890.9^b + 67.9^b = 959$ (-7)	958.6 (0)
		B_g	$\nu_{14} + \nu_{24}$	$717.9^b + 240.5^b = 958$ (-6)	947.2 (-11)
G, H	1083, 1105	A_g	$\nu_{10} + \nu_{12}$	$894^a + 210^a = 1104$	1100.2 (-1)
I	1238/1240/1244	A_g	$\nu_{21} + \nu_{24}$	$994.2^b + 240.5^b = 1235$	1232.9 (-2)
		A_g	$\nu_4 + \nu_9$	$1093^a + 157^a = 1250$	1269.3 (-4)
		A_g	$2\nu_7$	$2 \times 624^a = 1248$	1246.6 (-1)
J	1282	A_g	$\nu_4 + \nu_8$	$1093^a + 192^a = 1285$ (-3)	1302.9 (-4)
K	$1308/1319/1326/1355^c$	A_g	$\nu_{20} + \nu_{24}$	$1072.2^b + 240.5^b = 1313$	1309.7 (-3)
		A_g	$2\nu_{23}$	$2 \times 658.0^b = 1316$	1311.1 (-2)
		A_g	$2\nu_{11}$	$2 \times 668^a = 1336$	1320.9 (-13)
L	1406	A_g	$\nu_6 + \nu_9$	$1258^a + 157^a = 1415$ (-9)	1397.9 (-3)
		A_g	$2\nu_{14}$	$2 \times 717.9^b = 1436$ (-15)	1408.2 (-14)
M	1477	A_g	$\nu_{22} + \nu_{24}$	$1248.9^b + 240.5^b = 1489$ (-12)	1486.1 (-3)
N, O	1505/1509, 1537	A_g	?		
P	1556	A_g	$\nu_{10} + \nu_{11}$	$894^a + 668^a = 1562$ (-6)	1563.6 (0)
Q	1586	A_g	$\nu_{13} + \nu_{14}$	$890.9^b + 717.9^b = 1609$ (-23)	1608.5 (0)
R, S	1615, 1625	A_g	$\nu_5 + \nu_7$	$993^a + 624^a = 1617$	1626.9 (1)
T, U	1639, 1644/1647	A_g	$\nu_{21} + \nu_{23}$	$994.2^b + 658.0^b = 1652$	1651.2 (-1)
V	1656	B_g	$\nu_5 + \nu_{11}$	$993^a + 668^a = 1661$ (-5)	1674.3 (-1)
W, X	1684, 1707	A_g	$\nu_4 + \nu_7$	$1093^a + 624^a = 1717$	1741.7 (-5)
		B_g	$\nu_{14} + \nu_{21}$	$717.9^b + 994.2^b = 1712$	1709.9 (-2)
Y	1721	A_g	$\nu_{20} + \nu_{23}$	$1072.2^b + 658.0^b = 1730$ (-9)	1726.3 (-4)
Z	$1779/1784/1789^c$	A_g	$2\nu_{10}$	$2 \times 894^a = 1788$	1778.5 (-1)
		A_g	$2\nu_{13}$	$2 \times 890.9^b = 1782$	1780.3 (-1)
		A_g	$\nu_3 + \nu_9$	$1633^a + 157^a = 1790$	1770.0 (3)
		B_g	$\nu_{14} + \nu_{20}$	$717.9^b + 1072.2^b = 1790$	1784.8 (-5)

[a] Raman jet spectrum, this work.
[b] Anharmonic calculation results at the MP2/6-311+G(2d,p) level.
[c] Assumed Fermi resonance.

Table 3.17: Band positions (in cm^{-1}) and assignment of the Raman active overtone/combination bands of $(DCOOD)_2$ between 750 and 1800 cm^{-1}, compared with the anharmonic calculation at the MP2/6-311+G(2d,p) level (in cm^{-1}). The experimentally and theoretically derived "anharmonicity constants" of the combination bands $(x_{i,j})$ are given in parentheses. See text for more details.

Furthermore, in the Raman GP spectrum of $(DCOOD)_2$ [9], two bands at $1231.8/1781.1\,\mathrm{cm}^{-1}$ were observed and assigned as overtone bands of ν_7 and ν_{10}, respectively. Both of them are in agreement with the assignments in this work (see Tab. 3.17, band I and band Z). The Fermi-triad system in the IR-active C–O stretching region of $(HCOOH)_2$ has been discussed before. The jet-cooled IR spectra of the three isotopomers investigated with cavity ring-down spectroscopy [95] are simpler. Each of them shows one isolated main band (apparently unperturbed) and a weaker band nearby. In their Raman jet spectra, the C–O stretching bands ν_6 show mainly a single structure. The relatively strong combination band $\nu_{13} + \nu_{15}$ in the $(HCOOH)_2$ spectrum is also found in the jet spectrum of $(HCOOD)_2$, whereas $\nu_{21} + \nu_{24}$ has a similar wavenumber as ν_6, for both of the C-deuterated isotopomers.

It is interesting to see that the overtone bands of the Davydov pair of the intramolecular O–H/D out-of-plane bending mode, ν_{11} and ν_{14}, both have a more than $10\,\mathrm{cm}^{-1}$ theoretical anharmonicity constant (see Tabs. 3.16 and 3.17; the related bands of $(HCOOH)_2$ and $(DCOOH)_2$ can be found in Tabs. 3.18 and 3.22). The only exception is the overtone band of the ν_{14} mode of $(DCOOH)_2$ ($-1\,\mathrm{cm}^{-1}$, see Tab. 3.15). Overtone bands of the O–H/D in-plane bending modes have a even larger theoretical anharmonicity constant, which will be discussed in the next section. After all, the intramolecular O–H/D motions are much softer than the several intermolecular O–H/D. . . O modes in the low wavenumber region, whose overtones have very small anharmonicity effects.

No clear systematics is found during the assignment. Many combination bands with B_g symmetry are observed in the spectra of $(HCOOH)_2$, especially around the carbonyl stretching vibration, whereas the related bands of the deuterated isotopomers mainly have A_g symmetry. Systematic studies of concentration dependence will be needed to rule out larger clusters. The intensities of the same combination band of different compounds appear to be isotope dependent.

3.2.3 Region above 1800 cm^{-1}

Clarification of the complicated features in this wavenumber range is really a challenge, although some preparations are already made in the lower wavenumber range. Basically, the assignment will be made in the same way as in the preceding section, but with the help of IR jet spectroscopy, by comparison of the "corresponding" bands in the Raman and IR spectra.

At first, "corresponding" must be defined. The 24 fundamental modes of FAD correspond to 13 different kinds of vibrational motion. Beside the intermolecular twisting mode ν_{16} (only IR active) and the intermolecular stretching mode ν_8 (only Raman active), the other 11 vibrations occur in Davydov pairs of IR active and Raman active fundamentals, which are "corresponding" to each other. "Corresponding combination band" means that one of the components has been replaced by its Davydov partner (e.g., A_g to B_u). Therefore, a given combination band will have two corresponding bands. It is also possible that two combination bands assigned already in the Raman spectra have the same IR active correponding bands because these two combination bands are built up by two pairs of "corresponding fundamentals". The IR active corresponding band of a Raman active overtone band means the combination band between the related fundamental band and its "corresponding fundamental". Two overtone bands of a corresponding fundamental pair have only one corresponding IR active combination band.

Depolarization jet spectra of (HCOOH)$_2$ between 1800 and 2550 cm^{-1} are shown in Fig. 3.29. The band positions and assignments are listed in Tab. 3.18, also compared with the anharmonic prediction at the MP2/6-311+G(2d,p) level. The combination/overtone bands in this wavenumber range are rather weak because there is no fundamental transition nearby. Due to the low band intensities they are not easy to detect in GP spectra and therefore attracted little interest. Only a weak band at 1815.6 cm^{-1} was reported in the Raman GP spectrum [9] and assigned as the overtone of ν_{11}. This band corresponds to the band A or B in Fig. 3.29 with the same assignment.

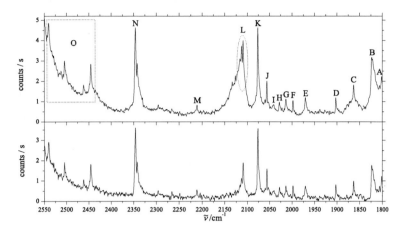

Figure 3.29: Depolarization analysis of $(HCOOH)_2$ between 1800 and 2550 cm^{-1}, with the excitation laser perpendicular to the scattering plane (top trace) and residual after subtracting 7/6 of the spectrum obtained with the excitation laser parallel to the scattering plane. $T_s = 16°C$ (3.7% HCOOH in He), $p_s = 700$ mbar, $d = 0.5$ mm, 8×200 s.

The strong combination bands in the C/O–H stretching vibration region begin near 2550 cm^{-1} and spread over about 600 cm^{-1} to 3200 cm^{-1}, with the intense ν_2 band at 2951 cm^{-1} in the centre. Depolarization jet spectra of $(HCOOH)_2$ between 2550 and 3300 cm^{-1} are shown in Fig. 3.30. The band positions and assignments are listed in Tab. 3.19. Many bands have been already observed and assigned in the Raman GP spectrum [9]. The assignments match ours completely except those for the several bands using ν_{20} as a combination partner (see Tab. 3.19). It is clear that the significantly different band positions of this mode ($\sim$1450 cm^{-1} in the IR GP spectra at room temperature [11,75] and 1406 cm^{-1} in the FTIR jet spectra in Fig. 3.19) used for the assignments lead to the different results.

The complicated band structure cannot be explained completely by the binary combinations listed in Tab. 3.19. Anharmonic coupling of the O–H stretching vibration with hydrogen bond vibration modes will contribute and combination

Label	Experiment	Symmetry	Assignment	Prediction	Calculation
A, B	1802/1807, 1824	A_g	$2\nu_{11}$	$2 \times 911^a = 1822$	$1782.6\ (-20)$
C	1863	A_g	$\nu_3 + \nu_9$	$1668^a + 161^a = 1829$	$1805.2\ (-5)$
		A_g	$\nu_3 + \nu_8$	$1668^a + 194^a = 1862\ (1)$	$1839.8\ (-3)$
		A_g	$2\nu_{14}$	$2 \times 939^b = 1878\ (-8)$	$1880.0\ (-10)$
D	1902	A_g	$\nu_6 + \nu_7$	$1224^a + 682^a = 1906\ (-4)$	$1895.7\ (-3)$
E	1970	A_g	$\nu_{10} + \nu_{11}$	$1059^a + 911^a = 1970\ (0)$	$1956.2\ (-11)$
F	1998	A_g	$\nu_{19} + \nu_{24}$	$1741^b + 264^b = 2005\ (-7)$	$1979.0\ (1)$
G	2014	A_g	$\nu_{13} + \nu_{14}$	$1069^b + 939^b = 2008\ (6)$	$1993.5\ (-25)$
H, I	2028, 2041	A_g	?		
J	2056	A_g	$\nu_5 + \nu_7$	$1376^a + 682^a = 2058\ (-2)$	$2065.6\ (-25)$
K	2076	A_g	$\nu_{21} + \nu_{23}$	$1373^b + 712^c = 2085\ (-9)$	$2079.3\ (0)$
L	2110/2113/2117	A_g	$2\nu_{10}$	$2 \times 1059^a = 2118$	$2108.4\ (-2)$
		A_g	$\nu_4 + \nu_7$	$1431^a + 682^a = 2113$	$2119.9\ (-3)$
		A_g	$\nu_{20} + \nu_{23}$	$1406^b + 712^c = 2118$	$2116.0\ (-4)$
M	2211	A_g	?		
N	2342/2346	A_g	$\nu_3 + \nu_7$	$1668^a + 682^a = 2350$	$2330.2\ (-7)$
O^d	2445/2462/2505/2540	A_g	$2\nu_6$	$2 \times 1224^a = 2448$	$2428.0\ (-5)$
		A_g	$\nu_{19} + \nu_{23}$	$1741^b + 712^c = 2453$	$2427.2\ (-3)$
		A_g	$2\nu_{22}$	$2 \times 1230.2^e = 2460$	$2440.6\ (-10)$

[a] Raman jet spectrum, this work.
[b] IR jet spectrum, courtesy of Kollipost.
[c] Halupka and Sander, Ar matrix isolation IR spectrum (Ref. 13).
[d] Assumed Fermi resonance.
[e] Ito, jet-cooled cavity ring-down IR spectra (Ref. 95).

Table 3.18: Band positions (in cm^{-1}) and assignment of the Raman active overtone/combination bands of $(HCOOH)_2$ between 1800 and 2550 cm^{-1} in Fig. 3.29, compared with the anharmonic calculation at the MP2/6-311+G(2d,p) level (in cm^{-1}). The experimentally and theoretically derived "anharmonicity constants" of the combination bands ($x_{i,i}$) are given in parentheses. See text for more details.

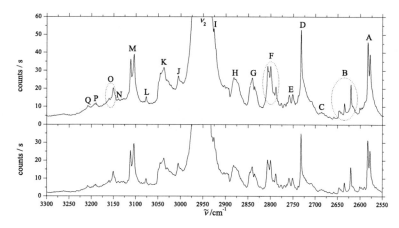

Figure 3.30: Depolarization analysis of $(HCOOH)_2$ between 2550 and 3300 cm^{-1}, with the excitation laser perpendicular to the scattering plane (top trace) and residual after subtracting 7/6 of the spectrum obtained with the excitation laser parallel to the scattering plane. $T_s = 16°C$ (3.7% HCOOH in He), $p_s = 700$ mbar, $d = 0.5$ mm, 8×200 s.

bands from more than two components (ternary overtone/combination bands) may exist as well. However, the binary combinations should be the dominant contributions, as was pointed out by the simulation analysis of the O–H stretching vibration of acetic acid dimer [17]. On the other side, the "corresponding" IR active bands of most of the Raman active bands assigned in Tab. 3.19 could be found in the IR jet spectra (see Fig. 3.31). All the FTIR jet spectra of FAD in this wavenumber region used in this work were recorded by P. Zielke [3,33], with a ∼0.3% compound concentration in He ($T_s = -20°C$). The band positions and assignments are listed in Tab. 3.20. The same labels but with lower case letters were used to mark these IR active bands compared to their Raman active "corresponding" bands. It was found that the wavenumber sequence of these labels in the IR spectrum is nearly the same as in the Raman case, but there is no direct relationship of the band intensity of the "corresponding" band partners.

Table 3.19: Band positions (in cm^{-1}) and assignments of the Raman active overtone/combination bands of (HCOOH)$_2$ between 2550 and 3300 cm^{-1} in Fig. 3.30, compared with the anharmonic calculation at the MP2/6-311+G(2d,p) level (in cm^{-1}). The experimentally and theoretically derived "anharmonicity constants" of the combination bands ($x_{i,j}$) are given in parentheses. See text for more details.

Label	Experiment Jet[a]	GP[b]	Symmetry	Assignment Jet[a]	Assignment GP[b]	Prediction	Calculation
A	2579/2583	2563	A_g	$\nu_5+\nu_6$	$\nu_5+\nu_6$	1376[a] + 1224[a] = 2600	2601.8 (−3)
B	2621/2635/2647	2616	A_g	$\nu_{21}+\nu_{22}$	$\nu_4+\nu_6$	1373[c] + 1230.2[d] = 2603	2591.8 (−13)
			A_g	$\nu_4+\nu_6$		1431[a] + 1224[a] = 2655	2646.4 (−15)
			A_g	$\nu_{20}+\nu_{22}$		1406[c] + 1230.2[d] = 2636	2627.8 (−16)
C	2686	2688	A_g	?			
D	2732	2726	A_g	$2\nu_{21}$	$2\nu_{21}$	2 × 1373[c] = 2746 (−7)	2744.0 (−3)
E	2751/2759	2746	A_g	$2\nu_5$	$2\nu_5$	2 × 1376[a] = 2752 (−4)	2762.9 (−4)
			A_g	$2\nu_{21}$		2 × 1373[c] = 2746 (−3)	2744.0 (−3)
F	2788/2800/2804/2807	2791	A_g	$\nu_{20}+\nu_{21}$	$\nu_{20}+\nu_{21}$	1407[c] + 1373[c] = 2779 (−13)	2775.8 (−13)
			A_g	$\nu_4+\nu_{21}$		1431[a] + 1376[a] = 2807 (−6)	2822.1 (−6)
G	2836/2841	2830	A_g	$2\nu_{20}$	$2\nu_{20}$	2 × 1406[c] = 2812 (−7)	2813.9 (−7)
H	2878/2882	2885	A_g	$2\nu_4$	$2\nu_4$	2 × 1431[a] = 2862	2823.4 (−31)
			A_g	$2\nu_4$	$2\nu_{20}$	2 × 1431[a] = 2862	2823.4 (−31)
I, J	2926, 3005		A_g	$\nu_3+\nu_6$	$\nu_3+\nu_5$	1668[a] + 1224[a] = 2892	2868.9 (−7)
			A_g	?		1668[a] + 1376[c] = 3044	3044.0 (−2)
K	3037/3045	3030	A_g	$\nu_3+\nu_5$	$\nu_3+\nu_4$	1668[a] + 1431[a] = 3099 (−22)	3099.2 (0)
L	3077	3084	A_g	$\nu_3+\nu_4$		1741[a] + 1373[c] = 3114	3107.7 (8)
M	3104/3111	3103	A_g	$\nu_{19}+\nu_{21}$	$\nu_{19}+\nu_{21}$	2951[a] + 161[a] = 3112	3111.4 (8)
N, O	3138, 3151/3160	3147	A_g	$\nu_2+\nu_9$	$\nu_{19}+\nu_{20}$	1728[c] + 1406[c] = 3134	3137.9 (1)
			A_g	$\nu_{19}+\nu_{20}$		2951[a] + 194[a] = 3145	3144.9 (10)
P	3190/3194	3186	A_g	$\nu_2+\nu_8$	$\nu_{18}+\nu_{20}$	2940[e] + 264[c] = 3204	3235.6 (19)
Q	3208		A_g	$\nu_{18}+\nu_{24}$		2940[e] + 264[c] = 3204 (4)	3235.6 (19)

[a] Raman jet spectrum, this work.
[b] Bertie and Michaelian, Raman GP spectrum (Ref. 9).
[c] IR jet spectrum, courtesy of Kollipost.
[d] Ito, jet-cooled cavity ring-down IR spectrum (Ref. 95).
[e] IR jet spectrum, courtesy of Zielke.

Label	Experiment				Assignment		Prediction	Calculation
	Jet[a]	Jet[b]	Jet[c]	GP[d]	This work	Ref.[b]		
a	2587		2586	2582	$\nu_5+\nu_{21}$ (B$_u$)		1224[e] + 1373[f] = 2597 (−10)	2587.8 (−6)
					$\nu_5+\nu_{22}$ (B$_u$)		1376[e] + 1230.2[g] = 2606	2607.2 (−8)
b	2624/2640/2646		2622/2525/2646	2623	$\nu_5+\nu_{20}$ (B$_u$)		1224[e] + 1406[f] = 2630	2623.7 (−9)
							911[e] + 1741[f] = 2652	2635.0 (0)
c	2657		2357		$\nu_{11}+\nu_{19}$ (A$_u$)		1431[e] + 1230.2[g] = 2661 (−4)	2653.8 (−19)
d/e	2726/2735/2741/2754		2729/2734/2767	2735	$\nu_4+\nu_{22}$ (B$_u$)		1668[e] + 1069[f] = 2737	2720.5 (−5)
					$\nu_3+\nu_{13}$ (A$_u$)		1376[e] + 1373[f] = 2749	2748.6 (−12)
f	2778/2792		2772/2792/2793		$\nu_5+\nu_{21}$ (B$_u$)		1376[e] + 1406[f] = 2782	2792.2 (−7)
	2801/2812		2814	2815	$\nu_5+\nu_{20}$ (B$_u$)		1431[e] + 1373[f] = 2804	2807.2 (−11)
	2825/2839	2825	2823/2824/2825/2839		$\nu_4+\nu_{21}$ (B$_u$)	$\nu_5+\nu_{20}$	1431[e] + 1406[f] = 2837	2832.9 (−23)
g	2877	2876	2878		$\nu_4+\nu_{20}$ (B$_u$)	$\nu_4+\nu_{20}$	1668[e] + 1230.2[g] = 2898 (−21)	2874.3 (−12)
h	2881/2896	2882	2883/2886/2897	2886	$\nu_3+\nu_{22}$ (B$_u$)	$\nu_3+\nu_{22}$[h]	1668[e] + 1230.2[g] = 2898 (18)	2874.3 (−12)
i	2916		2918	2916	$\nu_3+\nu_{22}$ (B$_u$)	$\nu_3+\nu_{22}$[h]	1224[e] + 1741[f] = 2965	2874.3 (−12)
h1	2946/2949	2949	2939/2949/2950	2957	$\nu_6+\nu_{19}$ (B$_u$)	$\nu_6+\nu_{19}$	1224[e] + 1741[f] = 2965 (9)	2939.1 (−3)
j/k	2974		2975		$\nu_2+\nu_{16}$ (A$_u$)	$\nu_6+\nu_{19}$	2951[e] + 69.2[i] = 3020	2939.1 (−3)
	2990/3005/3016/3029		2995/3016/3028	3028	$\nu_3+\nu_{21}$ (B$_u$)		1668[e] + 1373[f] = 3041 (8)	3023.5 (7)
	3049		3049					3028.8 (−3)
m	3067/3085/3094/3102	3081/95	3063/3084/3096/3103	3110	$\nu_3+\nu_{18}$ (B$_u$)	$\nu_{17}/(\nu_9+\nu_{18})$	161[e] + 2940[j] = 3101	3137.8 (22)
					$\nu_3+\nu_{20}$ (B$_u$)		1668[e] + 1406[f] = 3074	3060.7 (−10)
					$\nu_5+\nu_{19}$ (B$_u$)		1376[e] + 1741[f] = 3117	3113.2 (4)
l	3128	3128	3127		$\nu_2+\nu_{15}$ (A$_u$)		1376[e] + 1741[f] = 3117 (11)	3113.2 (4)
n	3139	3139	3139	3150	$\nu_8+\nu_{18}$ (B$_u$)		2951[e] + 168.5[i] = 3120 (−8)	3127.6 (9)
o	3157	3157	3158		$\nu_4+\nu_{19}$ (B$_u$)		194[e] + 2940[j] = 3134 (5)	3168.3 (20)
p/q	3175/3182/3203/3215		3180/3211	3210	$\nu_2+\nu_{24}$ (B$_u$)		1431[e] + 1741[f] = 3172 (−15)	3164.8 (−2)
q1	3241/3259/3268	3241/3259/3268	3257	3272	?[h]		2951[e] + 264[f] = 3215	3218.1 (14)

[a] IR jet spectrum, courtesy of Zielke.
[b] Georges et al., IR jet spectrum, 2800-3300 cm⁻¹ (Ref. 11).
[c] Raman jet spectrum, this work.
[d] Millikan and Pitzer, IR GP spectrum (Ref. 75).
[e] Ito, cavity ring-down IR jet spectrum (Ref. 95).
[f] IR jet spectrum, courtesy of Kollipost.
[g] Ito and Nakanaga, jet-cooled cavity ring-down IR spectrum (Ref. 108).
[h] Ternary combination bands are possible.
[i] Georges et al., high resolution FIR GP spectrum (Ref. 11).
[j] IR jet spectrum, courtesy of Zielke.

Table 3.20: Band positions (in cm⁻¹) and assignments of the IR active combination bands of (HCOOH)$_2$ between 2550 and 3300 cm⁻¹ in Fig. 3.31, compared with former measurements [11,75,109] and the anharmonic calculation at the MP2/6-311+G(2d,p) level (in cm⁻¹). The symmetries, the experimentally and theoretically derived "anharmonicity constants" of the combination bands ($x_{i,j}$) are given in parentheses. See text for more details.

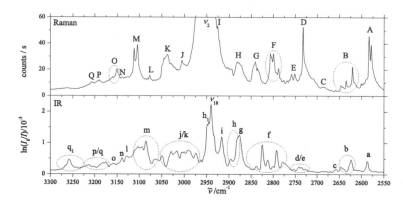

Figure 3.31: Comparison of the Raman/IR jet spectra of $(HCOOH)_2$ between 2550 and $3300\,\mathrm{cm}^{-1}$. The Raman jet spectrum is the same as the top trace in Fig. 3.30.

Other jet-cooled IR spectra of FAD in the O–H stretching region have been reported [11, 108, 109]. Several bands in the region of $2800\text{-}3300\,\mathrm{cm}^{-1}$ have been observed by cavity ring-down spectroscopy [108, 109]. From an analysis of the vibrational structure, most of them were assigned to binary combinations [108, 109]. Ternary combination bands were assumed for some bands but without detailed assignment due to the many possible candidates (normally more than five). The band positions and the assignments are included in Tab. 3.20. The band positions are in good agreement with ours, but the assignments using ν_{20} as a combination partner are different, due to the different band position of ν_{20} (around $1450\,\mathrm{cm}^{-1}$ in the IR GP spectra at room temperature [11, 75] and $1406\,\mathrm{cm}^{-1}$ in the FTIR jet spectra of this work) used for the assignments, like in the case of the Raman active combination bands in Tab. 3.19. The relatively strong band at $3085\,\mathrm{cm}^{-1}$ in our spectrum was previously observed at $3081\,\mathrm{cm}^{-1}$ [109] and assigned as the IR active O–H stretching fundamental ν_{17}. This assignment is relatively uncertain due to the large wavenumber difference between this experimental value and the anharmonic calculated wavenumber of $2990.4\,\mathrm{cm}^{-1}$ at the MP2/6-311+G(2d,p) level, as well as the extensive anharmonic couplings. The experimental result of the IR active C–H stretching fundamental vibration ν_{18} (2940 cm^{-1} in this work, 2939.7 cm^{-1} in Refs. 108,109) fits the theoretical value at the same calculation level quite well (2962.9 cm^{-1}). Furthermore, quantum chemical calculations [86] predict a very large vibrational intensity of the ν_{17} band (> 1500 km/mol), whereas the peak at 3081 cm^{-1} is much weaker than that of the C–H stretching band. The intensity of the ν_{17} band is considered to be shared among the manifold of combination bands through Fermi resonances [109].

R. Georges *et al.* investigated the IR spectrum of FAD in the range 2800-3300 cm^{-1} using jet-cooled laser spectroscopy [11]. The rotational temperature in the jet is estimated at 35 K, comparable to the typical translational/rotational temperature of 20 K of the filet-jet setup used for the measurements of our IR spectra [52]. Many bands were observed but no assignment was provided.

Millikan and Pitzer observed many combinations bands in the IR GP spectrum taken at 26°C but did not make the assignment either [75]. The band positions are listed in Tab. 3.20 too. Most of them are $\sim$10 cm^{-1} red shifted in comparison with those measured with low temperature techniques. Several bands from them [75] as well as from Georges *et al.* [11] which are out of the wavenumber range (below 2550 cm^{-1} or above 3300 cm^{-1}) of our IR jet spectrum in Fig. 3.31 are tentatively assigned in Tab. 3.21.

The bands in the Raman jet spectra of (DCOOH)$_2$ between 1800 and 2500 cm^{-1} (see Fig. 3.32) are assigned in Tab. 3.22. Two bands (L and M) observed in the Raman GP spectrum [67] are re-assigned in this work.

It should be noted that two nearby bands M and N (wavenumber difference: 9 cm^{-1}) have very high intensities, which are attributed to the Fermi resonance between the C–D stretching and two combination bands of A$_g$ symmetry ($\nu_5 + \nu_6$ and $\nu_{21} + \nu_{22}$). The corresponding IR active combination bands $\nu_{21} + \nu_6$ (marked with m in Fig. 3.33) and $\nu_5 + \nu_{22}$ (marked with n in Fig. 3.33) are detected in the IR jet spectra. Both bands have B$_u$ symmetries, relatively high intensities and very similar band positions as the Raman active ones after couplings, because there is almost no difference between the predicted wavenumbers of both corresponding

Jet[a]	GP[b]	Assignment	Prediction	Calculation
1935	1923	$\nu_3 + \nu_{24}$ (B$_u$)	$1668^c + 264^d = 1932$ (3)	1911.4 (1)
		$\nu_8 + \nu_{19}$ (B$_u$)	$194^c + 1741^d = 1935$ (0)	1909.7 (1)
		$\nu_6 + \nu_{23}$ (B$_u$)	$1224^c + 712^e = 1936$ (−1)	1920.6 (−5)
1992/1995		$\nu_{11} + \nu_{13}$ (B$_u$)	$911^c + 1069^d = 1980$	1954.5 (−25)
		$\nu_{12} + \nu_{19}$ (A$_u$)	$242^c + 1741^d = 1983$	1967.3 (−1)
		$\nu_{10} + \nu_{14}$ (B$_u$)	$1060^c + 939^d = 1999$	1998.8 (−7)
	2078	$\nu_7 + \nu_{20}$ (B$_u$)	$682^c + 1406^d = 2088$ (−10)	2092.5 (−2)
		$\nu_5 + \nu_{23}$ (B$_u$)	$1376^c + 712^e = 2088$ (−10)	2088.9 (−4)
2135		$\nu_{10} + \nu_{13}$ (B$_u$)	$1060^c + 1069^d = 2129$ (6)	2116.9 (−7)
		$\nu_{11} + \nu_{22}$ (A$_u$)	$911^c + 1230.2^f = 2141$ (−6)	2126.1 (−15)
		$\nu_4 + \nu_{23}$ (B$_u$)	$1431^c + 712^e = 2143$ (−8)	2143.6 (−6)
	2222	?		
2373		$\nu_4 + \nu_{14}$ (A$_u$)	$1431^c + 939^d = 2370$ (3)	2380.1 (−13)
		$\nu_3 + \nu_{23}$ (B$_u$)	$1668^c + 712^e = 2380$ (−7)	2357.8 (−6)
2435	2427	$\nu_7 + \nu_{19}$ (B$_u$)	$682^c + 1741^d = 2423$ (12)	2400.9 (−3)
		$\nu_{10} + \nu_{21}$ (A$_u$)	$1060^c + 1373^d = 2433$ (2)	2430.5 (0)
		$\nu_5 + \nu_{13}$ (A$_u$)	$1376^c + 1069^e = 2445$ (−10)	2453.1 (−1)
3392	3385	$\nu_3 + \nu_{19}$ (B$_u$)	$1668^e + 1741^d = 3409$ (−17)	3362.3 (−18)

[a] Georges et al., IR jet spectrum, 2800-3300 cm^{-1} (Ref. 11).

[b] Millikan and Pitzer, IR GP spectrum (Ref. 75).

[c] Raman jet spectrum, this work.

[d] IR jet spectrum, courtesy of Kollipost.

[e] Halupka and Sander, Ar matrix isolation IR spectrum (Ref. 13).

[f] Ito and Nakanaga, jet-cooled cavity ring-down IR spectrum (Ref. 108).

Table 3.21: Band positions (in cm^{-1}) and assignment of several IR active combination bands of (HCOOH)$_2$ from former studies [11, 75], compared with the anharmonic calculation at the MP2/6-311+G(2d,p) level (in cm^{-1}). The symmetries, the experimentally and theoretically derived "anharmonicity constants" of the combination bands ($x_{i,j}$) are given in parentheses.

IR/Raman combination band pairs. The comparison of the IR and Raman spectra of (DCOOH)$_2$ in the related wavenumber region is shown in Fig. 3.33. The band wavenumbers, assignment and the predictions are listed in Tab. 3.23.

The IR active bands in the jet spectrum in Fig. 3.33 are detected in the jet-cooled vibrational action spectrum [98] too and discussed in detail. A three-state deperturbation analysis shows that there is a relatively strong coupling between the IR active fundamental C–D stretching vibration ν_{18} and each of the combination vibrations (13 cm^{-1}) as well as between the combination states them-

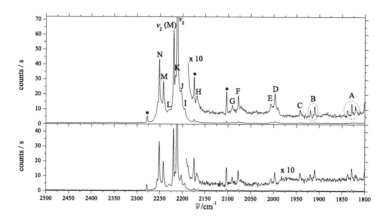

Figure 3.32: Raman depolarization analysis of $(DCOOH)_2$ between 1800 and 2500 cm^{-1}, with the excitation laser perpendicular to the scattering plane (top trace) and residual after subtracting 7/6 of the spectrum obtained with the excitation laser parallel to the scattering plane. $T_s = 0°C$ (1.4% DCOOH in He), $p_s = 700$ mbar, $d = 0.4$ mm, 8×150 s. The bands marked with * are from DCOOH monomer but not from fundamental modes. See Fig. 3.8 for more details.

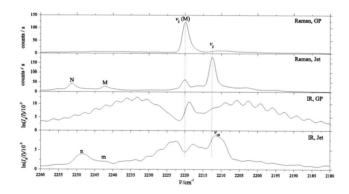

Figure 3.33: Comparison of the Raman/IR spectra of $(DCOOH)_2$ in the C–D stretching vibration region between 2180 and 2260 cm^{-1}. The Raman GP spectrum is the same as that in Fig. 3.8, the Raman jet spectrum is the same as the top trace in Fig. 3.32 and the two IR spectra were measured by P. Zielke. The labels in the Raman jet spectrum are the same as in the Fig. 3.32.

Label	Experiment Jet[a]	Experiment GP[b]	Assignment Symmetry	Assignment Jet[a]	Assignment GP[b]	Prediction	Calculation
A	1820/1829/1838		A_g	$\nu_{10}+\nu_{11}$		$894^a+921^a=1815$	$1803.6\ (-6)$
			A_g	$\nu_3+\nu_8$		$1643^a+193^a=1836$	$1812.6\ (-6)$
			A_g	$2\nu_{11}$		$2\times921^a=1842$	$1787.4\ (-24)$
B	1910/1919		A_g	$\nu_{13}+\nu_{14}$		$956/965^c+891/892^c=1847/1848/1856/1857$	$1849.7\ (-2)$
			A_g	$\nu_6+\nu_7$		$1238^a+676^a=1914$	$1898.4\ (-4)$
			A_g	$2\nu_{13}$		$2\times956/965^c=1912/1930$	$1879.3\ (-24)$
C	1942		A_g	$\nu_{22}+\nu_{23}$		$1244^c+700.6^d=1945\ (-3)$	$1932.9\ (-5)$
D, E	1997, 2006		A_g	$\nu_{19}+\nu_{24}$		$1708/1709/1723^c+248.7^d=957/1958/1972$	$1953.0\ (1)$
			A_g	$2\nu_5$		$2\times1001^a=2002$	$2018.7\ (-2)$
F	2077		A_g	$2\nu_{21}$		$2\times1003/1006^c=2006/2012$	$2025.6\ (-2)$
G	2091		A_g	$\nu_{20}+\nu_{23}$		$1381/1384^c+700.6^d=2082/2085$	$2073.0\ (-8)$
H	2168		A_g	$\nu_4+\nu_7$		$1418^a+676^a=2094$	$2108.2\ (-4)$
I	2195		A_g	?			
J	2203		A_g	?			
K	2216		A_g	?			
L	2230	2232	A_g	ν_2 (DCOOD)	$\nu_5+\nu_6$	$1001^a+1238^a=2239\ (-9)$	$2231.7\ (-7)$
M[e]	2242	2239	A_g	$2\nu_5$		$1001^a+1238^a=2239\ (3)$	$2231.7\ (-7)$
N[e]	2251		A_g	$\nu_{21}+\nu_{22}$	$\nu_{21}+\nu_{22}$	$1003/1006^c+1244^c=2247/2250$	$2261.1\ (-10)$

[a] Raman jet spectrum, this work.
[b] Bertie *et al.*, Raman GP spectrum (Ref. 67).
[c] Ito, Ar matrix isolation IR spectrum (Ref. 15).
[d] Anharmonic calculation results at the MP2/6-311+G(2d,p) level.
[e] Assumed Fermi resonance.

Table 3.22: Band positions (in cm^{-1}) and assignments of the Raman active overtone/combination bands of (DCOOH)$_2$ between 1800 and 2500 cm^{-1} in Fig. 3.32, compared with the anharmonic calculation at the MP2/6-311+G(2d,p) level (in cm^{-1}). The predictions of the assignment in Ref. 67 are given in italic face and the experimentally and theoretically derived "anharmonicity constants" of the combination bands ($x_{i,j}$) are given in parentheses. See text for more details.

| | Experiment | | | | |
Label	Jet	GP	Assignment	Prediction	Calculation
Raman					
	2212	2210	ν_2 (dimer)		2235.1
	2220	2220	ν_2 (monomer)		2226.5
M	2242		$\nu_5 + \nu_6$	$1001^a + 1238^a = 2239$	2231.7
N	2251/2255		$\nu_{21} + \nu_{22}$	$1003/06^b + 1244^b = 2247/50$	2261.1
FTIR					
	2211, 2210.4^c		ν_{18} (dimer)		2221.3
	2218	2219	ν_2 (monomer)		2226.5
m	2242, 2242.5^c		$\nu_{21} + \nu_6$	$1003/06^b + 1238^a = 2241/44$	2243.0
n	2248, 2248.3^c		$\nu_5 + \nu_{22}$	$1001^a + 1244^b = 2245$	2259.6

^{a} Raman jet spectrum, this work.

^{b} Ito, Ar matrix isolation IR spectrum (Ref. 15).

^{c} Yoon *et al.*, jet cooled vibrational action IR spectrum (Ref. 98).

Table 3.23: Band maxima (in cm^{-1}) and assignment of the Raman and IR active bands of DCOOH monomer and dimer in the spectra of Fig. 3.33, compared with the anharmonic calculation at the MP2/6-311+G(2d,p) level (in cm^{-1}). See text for more details.

selves ($7\,cm^{-1}$). The three bands of ν_{18}, m and n are observed at $2210.4\,cm^{-1}$, $2242.5\,cm^{-1}$ and $2248.3\,cm^{-1}$, whereas the related bands in our IR jet spectra are at $2211\,cm^{-1}$, $2242\,cm^{-1}$ and $2248\,cm^{-1}$.

Depolarization analysis of $(DCOOH)_2$ between 2500 and $3300\,cm^{-1}$ is shown in Fig. 3.34. The band positions and assignments are listed in Tab. 4.24, compared with the anharmonic calculation at the MP2/6-311+G(2d,p) level. The bands are weaker than those of $(HCOOH)_2$ in this wavenumber region, clearly due to the absence of the contribution of ν_2 by the couplings.

A comparison of the IR and Raman jet spectra between 2500 and $3300\,cm^{-1}$ is shown in Fig. 3.35. The IR band positions are listed in Tab. 3.25. Most of them can be assigned as the corresponding Raman active bands in Tab. 4.24. Nevertheless, the rough assignments cannot be concrete enough in detail but only as a reference for the many IR sub bands in Fig. 3.35.

Many "cold" IR active bands in this wavenumber range have been observed with cavity ring-down spectroscopy [109] and in rare gas matrices [15]. The related band

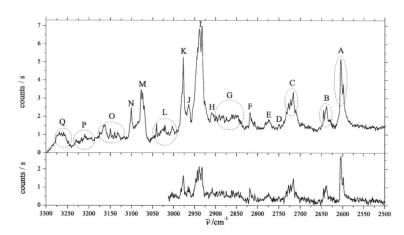

Figure 3.34: Raman depolarization analysis of $(DCOOH)_2$ between 2500 and $3300\,\mathrm{cm}^{-1}$, with the excitation laser perpendicular to the scattering plane (top trace) and residual after subtracting 7/6 of the spectrum obtained with the excitation laser parallel to the scattering plane. $T_s = 0°C$ (1.4% DCOOH in He), $p_s = 700\,\mathrm{mbar}$, $d = 0.4\,\mathrm{mm}$, $8 \times 150\,\mathrm{s}$.

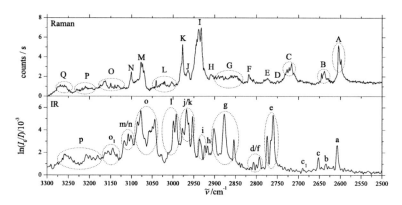

Figure 3.35: Comparison of the Raman/IR jet spectra of $(DCOOH)_2$ between 2500 and $3300\,\mathrm{cm}^{-1}$. The Raman jet spectrum is the same as the top trace in Fig. 3.34.

Label	Experiment Jet[a]	Experiment GP[b]	Symmetry	Assignment Jet[a]	Assignment GP[b]	Prediction	Calculation
A	2599/2604	2598	A_g	$\nu_{20}+\nu_{22}$	$\nu_4+\nu_6$	$1381/1384^c + 1244^a = 2625/2628$	2606.5 (−10)
B	2638/2645	2632	A_g	$\nu_3+\nu_5$	$\nu_3+\nu_5$	$1643^a + 1001^a = 2644$	2636.4 (−9)
			A_g	$\nu_4+\nu_6$		$1418^a + 1238^a = 2656$	2652.2 (−13)
C	2716/2722/2727	2700/2720	A_g	$\nu_{19}+\nu_{21}$	$\nu_{19}+\nu_{21}$	$1708/1709/1723^c + 1003/1006^c = 2711/\ldots/2729$	2716.1 (−2)
			A_g	$2\nu_{20}$	$2\nu_{20}$	*$2 \times 1381/1384^c = 2762/2768$*	*2731.0 (−14)*
D	2749	2745	A_g	$2\nu_{20}$		$2 \times 1381/1384^c = 2762/2768$	2731.0 (−14)
E	2774	2768	A_g	$2\nu_{20}$	$2\nu_4$	$2 \times 1381/1384^c = 2762/2768$	2731.0 (−14)
F	2819		A_g	$2\nu_4$		$2 \times 1418^a = 2836$ (−9)	2913.8 (−19)
G[d]	2840–2890	2840/2860	A_g	$\nu_3+\nu_6$	$\nu_3+\nu_6$	$1643^a + 1238^a = 2881$	2852.2 (−10)
			A_g	$\nu_2+\nu_7$		$2212^a + 676^a = 2888$	2920.8 (11)
H	2908		A_g	$\nu_{18}+\nu_{23}$		$2211^e + 700.6^c = 2912$ (−4)	2938.4 (17)
I	2933/2939/2943	2947	A_g	$\nu_{19}+\nu_{22}$	$\nu_{19}+\nu_{22}$	$1708/1709/1723^c + 1244^c = 2952/2953/2967$	2935.5 (−5)
J,K	2965, 2977	2965	A_g	?			
L	3003/3021/3041	3005	A_g	$\nu_3+\nu_4$	$\nu_3+\nu_4$	$1643^a + 1418^a = 3061$	3073.5 (1)
M	3070/3074/3077	3070/3080	A_g	$\nu_{19}+\nu_{20}$	$\nu_{19}+\nu_{20}$	$1708/1709/1723^c + 1381/1384^c = 3089/\ldots/3107$	3084.3 (1)
N	3100	3092	A_g	$\nu_{19}+\nu_{20}$		$1708/1709/1723^c + 1381/1384^c = 3089/\ldots/3107$	3084.3 (1)
O[d]	3133/3140/3150/3164	3130	B_g	$\nu_2+\nu_{11}$		$2212^a + 921^a = 3133$	3162.5 (9)
			B_g	$\nu_{13}+\nu_{18}$		$956/965^c + 2211^e = 3167/3176$	3199.3 (15)
P[d]	3194/3210/3217/3233	3232	A_g	$\nu_{18}+\nu_{21}$	$\nu_{18}+\nu_{21}$	$2211^e + 1003/1006^c = 3214/3217$	3244.0 (8)
			A_g	$\nu_2+\nu_5$		$2212^a + 1001^a = 3213$	3254.2 (8)
Q	3248/3259/3264/3269	3247	A_g	$2\nu_3$	$\nu_2+\nu_5$	$2 \times 1643^a = 3286$	3243.4 (−13)

[a] Raman jet spectrum, this work.
[b] Bertie *et al.*, Raman GP spectrum (Ref. 67).
[c] Ito, Ar matrix isolation IR spectrum (Ref. 15).
[d] Assumed Fermi resonance.
[e] IR jet spectrum, courtesy of Zielke.

Table 3.24: Band positions (in cm^{-1}) and assignments of the Raman active overtone/combination bands of $(DCOOH)_2$ between 2500 and 3300 cm^{-1} in Fig. 3.34, compared with the anharmonic calculation at the MP2/6-311+G(2d,p) level (in cm^{-1}). The predictions of the assignment in Ref. 67 are given in italic face and the experimentally and theoretically derived "anharmonicity constants" of the combination bands ($x_{i,j}$) are given in parentheses. See text for more details.

Label	Experiment				Assignment		Prediction	Calculation
	Jet		Ar Matrix	GP				
	This work	Ref. 109	Ref. 15	Ref. 75	This work	Ref. 109		
a	2608				$\nu_5 + \nu_{20}$ (B$_u$)	$\nu_5 + \nu_{20}$ (B$_u$)	1238[a] + 1379.9[b] = 2618 (−10)	2597.5 (−10)
b	2634				$\nu_3 + \nu_{13}$ (A$_u$)	$\nu_3 + \nu_{13}$ (A$_u$)	1643[a] + 963.3[b] = 2607 (1)	2593.9 (−4)
c	2652				$\nu_3 + \nu_{21}$ (B$_u$)	$\nu_3 + \nu_{21}$ (B$_u$)	1643[a] + 1014.4[b] = 2657 (−23)	2640.0 (−9)
c$_1$	2690			2626	$\nu_4 + \nu_{22}$ (B$_u$)	$\nu_4 + \nu_{22}$ (B$_u$)	1418[a] + 1236.9[b] = 2655 (−3)	2664.0 (−11)
d/f	2793/2800/2807/2813				$\nu_5 + \nu_{19}$ (B$_u$)	$\nu_5 + \nu_{19}$ (B$_u$)	1001[a] + 1703.2[b] = 2704 (−14)	2712.7 (−1)
e	2760/2766/2774				$\nu_4 + \nu_{20}$ (B$_u$)	?	1418[a] + 1379.9[b] = 2798	2776.4 (−41)
g	2854/2877	2877	2867		$\nu_3 + \nu_{22}$ (B$_u$)	$\nu_3 + \nu_{22}$	1643[a] + 1236.9[b] = 2880	2861.1 (−10)
h	2902	2903	2887	2889	$\nu_7 + \nu_{18}$ (B$_u$)		676[a] + 2211[c] = 2887 (15)	2912.4 (17)
	2917				$\nu_2 + \nu_{23}$ (B$_u$)	$\nu_2 + \nu_{23}$	2212[a] + 700.6[b] = 2913 (4)	2947.0 (11)
i	2924/2936				$\nu_6 + \nu_{19}$ (B$_u$)	$\nu_6 + \nu_{19}$	1238[a] + 1703.2[b] = 2941	2926.8 (−4)
j/k	2953/2962/2967/2974/2977	2969	2962	2962	$\nu_3 + \nu_{20}$ (B$_u$)	$\nu_2 + \nu_{20}$	1238[a] + 1703.2[b] = 2941	2926.8 (−4)
l	2992/2999/3012	2993	2989	3002	$\nu_4 + \nu_{19}$ (B$_u$)	$\nu_3 + \nu_{20}$	1643[a] + 1379.9[b] = 3023	2998.0 (−16)
m/n	3101/3108/3117	3100		3098	$\nu_2 + \nu_{14}$ (A$_u$)	$\nu_4 + \nu_{19}$	1418[a] + 1703.2[b] = 3121	3140.8 (0)
o[d]	3044/3050/3056/3078/3086	3045/3078	3036/3069		$\nu_{10} + \nu_{18}$ (A$_u$)	ternary/ν_{17}	2212[a] + 888.1[b] = 3100	3130.9 (8)
o$_1$[d]	3135/3145/3156/3163				$\nu_{11} + \nu_{18}$ (A$_u$)		894[a] + 2211[c] = 3105	3122.1 (8)
o$_1$[d]	3181/3191/3199/3208			3173	$\nu_2 + \nu_{13}$ (A$_u$)		921[a] + 2211[c] = 3132	3154.2 (15)
p[d]	3245/3259			3270	$\nu_3 + \nu_{18}$ (B$_u$)		2212[a] + 963.3[b] = 3175 2212[a] + 1014.4[b] = 3216 1001[a] + 2221.3[b] = 3221	3207.6 (9) 3252.5 (3) 3240.3 (8)

[a] Raman jet spectrum, this work.
[b] Anharmonic calculation results at the MP2/6-311+G(2d,p) level.
[c] IR jet spectrum, courtesy of Zielke.
[d] Assumed Fermi resonance.

Table 3.25: Band positions (in cm^{-1}) and assignments of the IR active combination bands of (DCOOH)$_2$ between 2500 and 3300 cm^{-1} in Fig. 3.35, compared with the anharmonic calculation at the MP2/6-311+G(2d,p) level (in cm^{-1}). The symmetries, the experimentally and theoretically derived "anharmonicity constants" of the combination bands ($x_{i,j}$) are given in parentheses. See text for more details. Due to the space limit of the table the anharmonic calculation results at the MP2/6-311+G(2d,p) level are used for the prediction, instead of the band positions of the Ar-matrix experiment [15]).

positions and assignments are also listed in Tab. 3.25. All the assignments are in good agreement with these in this work. The relatively strong band at $3078\,\mathrm{cm^{-1}}$ in our spectrum was observed at the same position and assigned as the IR active O–H stretching fundamental ν_{17} by Ref. 109. Just like in the case of $(HCOOH)_2$, the experimental value does not fit well to the anharmonic calculated wavenumber at the MP2/6-311+G(2d,p) level ($2979.1\,\mathrm{cm^{-1}}$).

The band positions from the former IR GP study [75] are listed in Tab. 3.25 for comparison. Several bands below $2500\,\mathrm{cm^{-1}}$ are assigned in Tab. 3.26.

Experiment	Assignment	Prediction	Calculation
1891	$\nu_3 + \nu_{24}$ (B_u)	$1643^a + 248.7^b = 1892\ (-1)$	$1879.4\ (-4)$
	$\nu_5 + \nu_{14}$ (A_u)	$1001^a + 891/892^c = 1892/1893$	$1899.3\ (0)$
	$\nu_{10} + \nu_{21}$ (A_u)	$894^a + 1003/1006^c = 1897/1900$	$1902.4\ (-3)$
	$\nu_8 + \nu_{19}$ (B_u)	$193^a + 1708/1709/1723^c = 1901/1902/1916$	$1887.7\ (1)$
	$\nu_{12} + \nu_{19}$ (A_u)	$212^a + 1708/1709/1723^c = 1920/1921/1935$	$1916.3\ (-1)$
	$\nu_7 + \nu_{22}$ (B_u)	$676^a + 1244^c = 1920\ (-29)$	$1907.9\ (-3)$
2064	$\nu_7 + \nu_{20}$ (B_u)	$676^a + 1381/1384^c = 2057/2060$	$2051.3\ (-3)$
2153	$\nu_4 + \nu_{23}$ (B_u)	$1418^a + 700.6^b = 2119\ (34)$	$2131.1\ (-7)$
	$\nu_6 + \nu_{14}$ (A_u)	$1238^a + 891/892^c = 2129/2130$	$2114.2\ (-2)$
	$\nu_{10} + \nu_{22}$ (A_u)	$894^a + 1244^c = 2138\ (15)$	$2123.0\ (-5)$
	$\nu_{11} + \nu_{22}$ (A_u)	$921^a + 1244^c = 2165\ (-12)$	$2146.8\ (-8)$
2224	$\nu_6 + \nu_{21}$ (B_u)	$1238^a + 1003/1006^c = 2241/2244$	$2243.0\ (1)$
	$\nu_5 + \nu_{22}$ (B_u)	$1001^a + 1244^c = 2245\ (-21)$	$2259.6\ (12)$
2251	$\nu_5 + \nu_{22}$ (B_u)	$1001^a + 1244^c = 2245\ (6)$	$2259.6\ (12)$
	$\nu_{10} + \nu_{20}$ (A_u)	$894^a + 1381/1384^c = 2275/2278$	$2265.2\ (-6)$
	$\nu_2 + \nu_{16}$ (A_u)	$2212^a + 66.9^b = 2279\ (-28)$	$2313.5\ (12)$
2316	$\nu_{11} + \nu_{20}$ (A_u)	$921^a + 1381/1384^c = 2302/2305$	$2282.4\ (-16)$
	$\nu_4 + \nu_{14}$ (A_u)	$1418^a + 891/892^c = 2309/2310$	$2323.3\ (-3)$
	$\nu_3 + \nu_{23}$ (B_u)	$1643^a + 700.6^b = 2344\ (-28)$	$2326.5\ (-9)$
2440	$\nu_{12} + \nu_{18}$ (A_u)	$212^a + 2211^d = 2423\ (17)$	$2451.0\ (16)$
	$\nu_2 + \nu_{24}$ (A_u)	$2212^a + 248.7^b = 2461\ (-21)$	$2494.7\ (11)$

[a] Raman jet spectrum, this work.
[b] Anharmonic calculation results at the MP2/6-311+G(2d,p) level.
[c] Ito, Ar matrix isolation IR spectrum (Ref. 15).
[d] IR jet spectrum, courtesy of Zielke.

Table 3.26: Band positions (in $\mathrm{cm^{-1}}$) and assignment of several IR active combination bands of $(DCOOH)_2$ observed in former IR GP spectrum [75], compared with the anharmonic calculation at the MP2/6-311+G(2d,p) level (in $\mathrm{cm^{-1}}$). The symmetries, the experimentally and theoretically derived "anharmonicity constants" of the combination bands ($x_{i,j}$) are given in parentheses.

Depolarization analysis of $(HCOOD)_2$ between 1800 and 2525 cm^{-1} is shown in Fig. 3.36. The band positions and assignments are listed in Tab. 3.27. Most bands have been observed and assigned in the Raman GP spectra [67] (see Tab. 3.27), but the binary combination bands are mainly reassigned, based on the use of more accurate fundamental band positions. Some bands were assigned in Ref. 67 as ternary or even quaternary combination bands, and these kinds of combinations are not discussed in this work because of the many combination possibilities.

Comparison of the IR and Raman jet spectra in the same wavenumber range is shown in Fig. 3.37. Band positions of the IR active bands are listed in Tab. 3.28, compared with the former GP experimental values [75]. Several bands below 2000 cm^{-1} were observed in the former IR GP study [75] and they are assigned in Tab. 3.29. The intense band i has a similar wavenumber as the related Raman active one and remains without binary combination assignment. The relatively strong band at 2200 cm^{-1} in the Raman jet spectrum (marked with I) could be the O–D stretching fundamental ν_1 [33]. This prediction is relatively uncertain (marked with "?" in Tab. 3.27) due to the its small band intensity as well as the relatively large wavenumber difference from the anharmonic calculation at the MP2/6-311+G(2d,p) level (2158.0 cm^{-1}). The intense band in the IR jet spectrum (marked with j) at 2236 cm^{-1} could be the related IR active fundamental ν_{17}. This wavenumber fits the anharmonic calculated value (2256 cm^{-1}) at the MP2/6-311+G(2d,p) level quite well. Both O–D stretching fundamentals are used as combination components for the band assignments listed in Tabs. 3.27 and 3.28. The "anharmonicity constants" are satisfactorily small, supporting the assumption of the two fundamental band positions. All the O/C–H/D stretching fundamentals of FAD are summarized in Tab. 3.35 and will be discussed in detail later on.

For these unassigned bands around the O–D stretching fundamentals, both IR and Raman active, a significant regularity of the band distance is observed [33], compared to the related spectra of $(HCOOH)_2$ and $(DCOOH)_2$ in the O–H stretching region. A clear interpretation is still missing.

No depolarization experiment is carried out for $(HCOOD)_2$ above 2525 cm^{-1}. The Raman jet spectrum between 2525 and 3300 cm^{-1} is directly compared with

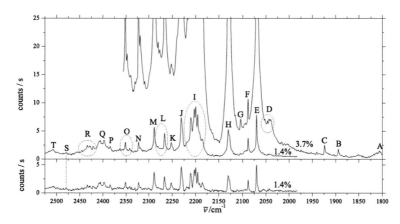

Figure 3.36: Raman depolarization analysis of (HCOOD)$_2$ between 1800 and 2525 cm^{-1}, with the excitation laser perpendicular to the scattering plane (top trace) and residual after subtracting 7/6 of the spectrum obtained with the excitation laser parallel to the scattering plane. $T_s = 0°$C (1.4% HCOOD in He), $p_s = 700$ mbar, $d = 0.4$ mm, 8×100 s. The spectrum with 3.7% HCOOD in He in top trace is the same as that in Fig. 3.9.

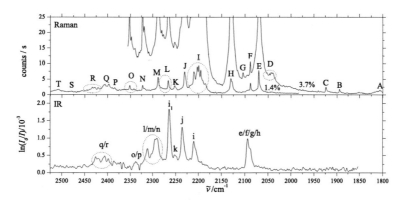

Figure 3.37: Comparison of the Raman/IR jet spectra of (HCOOD)$_2$ between 1800 and 2525 cm^{-1}. The Raman jet spectra are the same as those in the top trace of Fig. 3.36.

Table 3.27: Band positions (in cm^{-1}) and assignments of the Raman active overtone/combination bands of (HCOOD)$_2$ between 1800 and 2525 cm^{-1} in Fig. 3.36, compared with the anharmonic calculation at the MP2/6-311+G(2d,p) level (in cm^{-1}). The predictions of the assignment in Ref. 67 are given in italic face and the experimentally and theoretically derived "anharmonicity constants" of the combination bands ($x_{i,j}$) are given in parentheses. See text for more details.

Label	Experiment Jet[a]	Experiment GP[b]	Symmetry	Assignment Jet[a]	Assignment GP[b]	Prediction	Calculation
A	1807		A_g	$\nu_3+\nu_9$		$1653^a + 157^a = 1810\ (-3)$	1787.4 (1)
B	1894		A_g	$\nu_6+\nu_7$		$1268^a + 628^a = 1896\ (-2)$	1889.5 (−3)
C	1924		A_g	$\nu_{22}+\nu_{23}$		$1260.9^c + 662.8^c = 1924\ (0)$	1920.2 (−4)
D	2043/2046/2052	2047	A_g	$\nu_{21}+\nu_{23}$	$\nu_{21}+\nu_{23}$	$1399.9^c + 662.8^c = 2063$	2062.4 (0)
E[d], F[d], G	2070, 2088, 2104	2004, ..., 2097	A_g	$2\nu_{20}$ / $2\nu_{13}$ / $2\nu_{10}$	$2\nu_{20}$ / $2\nu_{13}$ / $2\nu_{10}$	$2 \times 1056.4^a = 2113$ / $2 \times 1055.5^c = 2112$ / $2 \times 1057^a = 2114$	2080.4 (−16) / 2108.8 (−1) / 2101.6 (−1)
H	2130	2124	A_g	$2\nu_{10}$	$2\nu_{10}$	$2 \times 1057^a = 2114\ (8)$	2242.6 (24)
I	2185/2195/2200/2203, 2210/2219, 2231	2200, 2210, 2231	A_g	$\nu_1?$, ?, $2\nu_4$	$\nu_4+2\nu_7$, $\nu_4+2\nu_{20}$, $\nu_9+2\nu_{20}$	$1076^a + 2 \times 628^a = 2332$ / $157^a + 2 \times 1056.4^c = 2270$ / $1076^b + 126^b = 2344$	2354.7 (−3) / 2314.6 (−3) / 2354.7 (−6)
J	2231	2231	A_g	?	$\nu_8+2\nu_{20}$	$194^a + 2 \times 1056.4^c = 2307$	2342.5 (34)
K	2252	2255	A_g	?			
L	2266/2277/2280	2274	A_g	$\nu_3+\nu_7$	$\nu_3+\nu_7$	$1653^a + 628^a = 2281$	2267.6 (3)
M	2288	2286	A_g	$\nu_{20}+\nu_{22}$	$\nu_{20}+\nu_{22}$	$1056.4^a + 1260.9^c = 2317$	2267.6 (3)
N	2319/2323	2300	A_g	$\nu_{20}+\nu_{22}$	$\nu_{18}+\nu_{24}$	$1056.4^a + 1260.9^c = 2317$	2314.6 (−3)
O	2340/2348/2351	2345	A_g	$\nu_4+\nu_6$	$\nu_4+\nu_5$	$1076^a + 1268^a = 2344$ / $2200^b + 157^a = 2357$	2354.7 (−6)
P	2384	2380	A_g	$\nu_{19}+\nu_{23}$	$\nu_{19}+\nu_{23}$	$1719.2^c + 662.8^c = 2382\ (2)$	2378.6 (−3)
Q	2398/2405	2413	A_g	$\nu_1+\nu_8?$	$\nu_{20}+\nu_{21}$	$2200^b? + 194^a = 2394$	2385.7 (41)
R[d]	2422/2428/2434/2446	2446	A_g	$\nu_{20}+\nu_{21}$ / $\nu_4+\nu_5$	$\nu_6+2\nu_7$ / $4\nu_7$	$1056.4^a + 1399.9^c = 2456$ / $1076^b + 1382^b = 2458$ / $2236^c + 245.7^c = 2482\ (-2)$	2455.7 (−1) / 2490.3 (−1) / 2483.6 (−18)
S	2480		A_g	$2\nu_{22}$	$2\nu_{22}$	$2 \times 1260.9^a = 2522\ (-8)$	2514.3 (−4)
T	2507		A_g	$2\nu_6$	$2\nu_6$	$2 \times 1268^a = 2536\ (-15)$	2518.6 (−4)

[a] Raman jet spectrum, this work.
[b] Bertie et al., Raman GP spectrum (Ref. 67).
[c] Anharmonic calculation results at the MP2/6-311+G(2d,p) level.
[d] Assumed Fermi resonance.
[e] IR jet spectrum, courtesy of Zielke.

Table 3.28: Band positions (in cm^{-1}) and assignment of the IR active combination bands of (HCOOD)$_2$ between 1800 and 2525 cm^{-1} in Fig. 3.37, compared with the former GP study [75] as well as the anharmonic calculation at the MP2/6-311+G(2d,p) level (in cm^{-1}). Several bands observed only in the GP spectrum [75] are also assigned. The symmetries, the experimentally and theoretically derived "anharmonicity constants" of the combination bands ($x_{i,j}$) are given in parentheses. See text for more details.

Label	Jet[a]	GP[b]	Assignment	Prediction	Calculation
e/f/g/h	2089/2094	2080	$\nu_{10} + \nu_{13}$ (B_u)	1057[c] + 1055.8[d] = 2113	2102.5 (−6)
			$\nu_4 + \nu_{20}$ (B_u)	1076[c] + 1056.4[d] = 2132	2104.2 (−49)
			$\nu_{10} + \nu_{20}$ (A_u)	1057[c] + 1055.8[d] = 2113	2102.5 (−6)
			$\nu_5 + \nu_{14}$ (A_u)	1382[c] + 723.2[d] = 2105	2116.8 (0)
			$\nu_4 + \nu_{13}$ (A_u)	1076[c] + 1055.8[d] = 2132	2151.5 (−2)
		2162	$\nu_4 + \nu_{13}$ (A_u)	1076[c] + 1055.8[d] = 2132 (30)	2151.5 (−2)
i	2210/2218		?		2256.3
j	2236		ν_{17} (B_u)	2200[c]? + 68.1[d] = 2268 (−16)	2255.9 (30)
k	2252		$\nu_1 + \nu_{16}$ (A_u)	2200[c]? + 68.1[d] = 2268 (−4)	2255.9 (30)
i$_1$	2264	2263	$\nu_1 + \nu_{16}$ (A_u)	2200[c]? + 68.1[d] = 2268	2255.9 (30)
l/m/n	2291/2295/2303/2311	2314	$\nu_1 + \nu_{16}$ (A_u)	1653[c] + 662.8[d] = 2316	2302.0 (3)
			$\nu_3 + \nu_{23}$ (B_u)	1268[c] + 1056.4[d] = 2324	2315.5 (−4)
			$\nu_6 + \nu_{20}$ (B_u)	1076[c] + 1260.9[d] = 2337 (1)	2356.3 (−2)
o/p	2338		$\nu_4 + \nu_{22}$ (B_u)	628[c] + 1719.2[d] = 2347 (−9)	2346.1 (−2)
			$\nu_7 + \nu_{19}$ (B_u)	2200[c]? + 162.2[d] = 2362 (−24)	2360.2 (40)
			$\nu_1 + \nu_{15}$ (A_u)	2200[c]? + 162.2[d] = 2362 (8)	2360.2 (40)
		2370	$\nu_1 + \nu_{15}$ (A_u)	1653[c] + 723.2[d] = 2376 (−6)	2372.5 (13)
			$\nu_3 + \nu_{14}$ (A_u)	674[c] + 1719.2[d] = 2393	2395.7 (1)
q/r	2398/2406/2420/2426	2455	$\nu_{11} + \nu_{19}$ (A_u)	157[c] + 2236[a] = 2393	2369.9 (−37)
			$\nu_9 + \nu_{17}$ (B_u)	194[c] + 2236[a] = 2430	2411.6 (−31)
			$\nu_8 + \nu_{17}$ (B_u)	1382[c] + 1056.4[d] = 2438	2449.6 (0)
			$\nu_5 + \nu_{20}$ (B_u)	2200[c]? + 245.7[d] = 2446	2462.6 (59)
			$\nu_1 + \nu_{24}$ (B_u)	238[c] + 2236[a] = 2474	2468.3 (−29)
			$\nu_{12} + \nu_{17}$ (A_u)	1076[c] + 1399.9[d] = 2476	2496.4 (−1)
			$\nu_4 + \nu_{21}$ (B_u)		

[a] IR jet spectrum, courtesy of Zielke.
[b] Millikan and Pitzer, IR GP spectrum (Ref. 75).
[c] Raman jet spectrum, this work.
[d] Anharmonic calculation results at the MP2/6-311+G(2d,p) level.

Experiment	Assignment	Prediction	Calculation
1605	$\nu_8 + \nu_{21}$ (B$_u$)	$194^a + 1399.9^b = 1594$ (11)	1586.1 (0)
	$\nu_5 + \nu_{24}$ (B$_u$)	$1382^a + 245.7^b = 1628$ (−23)	1639.1 (0)
1660	$\nu_{12} + \nu_{21}$ (A$_u$)	$238^a + 1399.9^b = 1638$ (32)	1640.5 (−1)
	$\nu_7 + \nu_{13}$ (A$_u$)	$628^a + 1055.8^b = 1684$ (−24)	1683.9 (−1)
1914	$\nu_7 + \nu_{22}$ (B$_u$)	$628^a + 1260.9^b = 1889$ (25)	1886.9 (−3)
	$\nu_3 + \nu_{24}$ (B$_u$)	$1653^a + 245.7^b = 1899$ (15)	1888.7 (7)
	$\nu_8 + \nu_{19}$ (B$_u$)	$194^a + 1719.2^b = 1913$ (1)	1906.2 (1)
	$\nu_6 + \nu_{23}$ (B$_u$)	$1268^a + 662.8^b = 1931$ (−17)	1921.1 (−5)
	$\nu_{11} + \nu_{22}$ (A$_u$)	$674^a + 1260.9^b = 1935$ (−21)	1931.0 (−53)

^{a} Raman jet spectrum, this work.

^{b} Anharmonic calculation results at the MP2/6-311+G(2d,p) level.

Table 3.29: Band positions (in cm^{-1}) and assignment of several IR active combination bands of (HCOOD)$_2$ observed in the former GP study [75], compared with the anharmonic calculation at the MP2/6-311+G(2d,p) level (in cm^{-1}). The symmetries, the experimentally and theoretically derived "anharmonicity constants" of the combination bands ($x_{i,j}$) are given in parentheses. See text for more details.

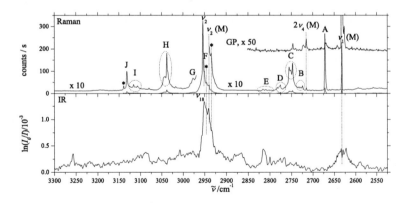

Figure 3.38: Comparison of the Raman/IR jet spectra of (HCOOD)$_2$ between 2525 and 3300 cm^{-1}. The Raman spectrum was recorded under the following conditions: $T_s = 0°C$ (1.4% HCOOD in He), $p_s = 700$ mbar, $d = 0.4$ mm, 8×200 s. A Raman GP spectrum taken from Fig. 3.9 is also shown in top trace for comparison. The bands marked with * are from HCOOD monomer but not from fundamental modes. See Fig. 3.9 for more details.

the IR jet spectrum (see Fig. 3.38). No clear IR active combination bands are observed due to a relatively poor signal/noise ratio in the spectrum. The most intense band of a band group at 2951 cm^{-1} is assigned as the fundamental IR active C–H stretching vibration mode ν_{18}. A related band was reported at 2936.2 cm^{-1} in the IR jet cavity ring-down studies of Ito *et al.* [109] as well as at 2960 cm^{-1} in the GP spectrum [75]. Compared to the ν_{18} band of (HCOOH)$_2$ at 2940 cm^{-1}, the band position of 2951 cm^{-1} shows a uniform agreement with the quantum chemical calculations. The anharmonic calculation at the MP2/6-311+G(2d,p) level provides a wavenumber of 2962.9 cm^{-1} for ν_{18} of (HCOOH)$_2$ as well as 2971.6 cm^{-1} for that of (HCOOD)$_2$. The calculated isotope shift (9 cm^{-1}) fits our experimental value (11 cm^{-1}) quite well. Another weak band was observed at 2572 cm^{-1} in the GP spectrum [75] but no suitable assignment can be found.

Wavenumbers and assignments of the Raman active bands are listed in Tab. 3.30. The bands observed and assigned in the Raman GP spectra [67] are listed for comparison. All the assignments except band M are same as these in this work.

The C/O–D stretching fundamental modes are all below 2300 cm^{-1}. Therefore, the spectra of the fully deuterated FAD are measured only up to 2450 cm^{-1}. Fig. 3.39 shows the depolarization jet spectra of (DCOOD)$_2$ between 1800 and 2450 cm^{-1}. The band positions and assignments are listed in Tab. 3.31, compared with the experimental results of former Raman GP measurements [9]. Band H can be distinguished from the C–D stretching fundamental mode ν_2 only by measurements with very low acid concentration. The relatively strong band H at 2215 cm^{-1} could contain a particularly high fraction of the O–D stretching fundamental mode ν_1.

Comparison of the IR and Raman jet spectra in the same wavenumber range is shown in Fig. 3.40 and the band positions and assignments are listed in Tab. 3.32. One of the two strongest bands at 2211 cm^{-1} is assigned as the IR active C–D stretching fundamental mode ν_{18}, with the completely same band position of ν_{18} of (DCOOH)$_2$. The other strong band at 2241 cm^{-1} could be the IR active O–D stretching fundamental mode ν_{17} (marked with j). Several bands below 1950 cm^{-1}

Table 3.30: Band positions (in cm^{-1}) and assignments of the Raman active overtone/combination bands of (HCOOD)$_2$ between 2525 and 3300 cm^{-1} in Fig. 3.38, compared with the anharmonic calculation at the MP2/6-311+G(2d,p) level (in cm^{-1}). The predictions of the assignment in Ref. 67 are given in italic face and the experimentally and theoretically derived "anharmonicity constants" of the combination bands ($x_{i,j}$) are given in parentheses. See text for more details.

| | Experiment | | | Assignment | | | |
Label	Jet[a]	GP[b]	Symmetry	Jet[a]	GP[b]	Prediction	Calculation
A	2672		B$_g$	$\nu_3 + \nu_{10}$		$1653^a + 1057^a = 2710\,(-38)$	2685.9 (−2)
			A$_g$	$\nu_{21} + \nu_{22}$		$1399.9^a + 1260.9^a = 2661\,(11)$	2659.3 (−2)
B	2724/2728/2733		A$_g$	$\nu_3 + \nu_4$		$1653^a + 1076^a = 2729$	2734.0 (1)
C[d]	2748/2752/2756	2750	A$_g$	$2\nu_5$	$2\nu_5$	$2 \times 1382^a = 2764$	2778.7 (−4)
			A$_g$	$\nu_{19} + \nu_{20}$		$1719.2^c + 1056.4^a = 2776$	2773.8 (−2)
D	2767/2776/2778/2783	2770	A$_g$	$2\nu_2$	$2\nu_{21}$	$2 \times 1399.9^a = 2800$	2792.3 (−4)
E	2802/2809/2816		A$_g$	$\nu_1 + \nu_7$		$2200^a + 628^a = 2828$	2823.0 (36)
			B$_g$	$\nu_{14} + \nu_{17}$		$723.2^c + 2236^e = 2959\,(8)$	2828
F	2951		B$_g$				2972.2 (−7)
G	2977		A$_g$	$\nu_3 + \nu_5$	$\nu_3 + \nu_5$	$1719.2^c + 1260.9^c = 2980\,(3)$	2977.4 (−3)
H	3038/3044	3044	A$_g$	$\nu_{19} + \nu_{22}$		$1653^a + 1382^a = 3035$	3039.9 (10)
			B$_g$	$\nu_{16} + \nu_{18}$		$68.1^c + 2951^e = 3019$	3058.9 (19)
			B$_g$	$\nu_2 + \nu_9$		$2954^a + 157^a = 3111$	3142.0 (21)
			A$_g$	$\nu_{15} + \nu_{18}$		$162.2^c + 2951^e = 3113$	3152.7 (19)
I[d], J	3107/3111/3116/3124, 3132	3128	A$_g$	$\nu_{19} + \nu_{21}$	$\nu_{19} + \nu_{21}$	$1719.2^c + 1399.9^c = 3119$	3123.0 (4)
			A$_g$	$\nu_2 + \nu_8$		$2954^a + 194^a = 3148$	3177.7 (21)

[a] Raman jet spectrum, this work.
[b] Bertie et al., Raman GP spectrum (Ref. 67).
[c] Anharmonic calculation results at the MP2/6-311+G(2d,p) level.
[d] Assumed Fermi resonance.
[e] IR jet spectrum, courtesy of Zielke.

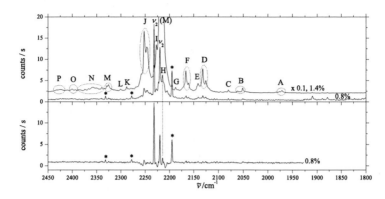

Figure 3.39: Raman depolarization analysis of (DCOOD)$_2$ between 1800 and 2450 cm^{-1}, with the excitation laser perpendicular to the scattering plane (top trace) and residual after subtracting 7/6 of the spectrum obtained with the excitation laser parallel to the scattering plane. $T_s = -8°C$ (0.8% DCOOD in He), $p_s = 700$ mbar, $d = 0.4$ mm, 8×150 s. Due to the low substance amount in the saturator, the signal intensity is much lower than that in the jet spectrum shown in Fig. 3.10 carried out under the same conditions. A Raman jet spectrum taken from Fig. 3.10 with 1.4% DCOOD in He is shown in top trace for comparison. The bands marked with * are from DCOOD monomer but not from fundamental modes. See Fig. 3.10 for more details.

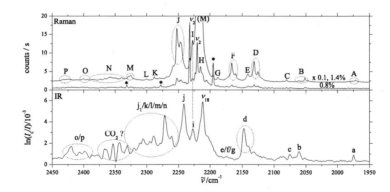

Figure 3.40: Comparison of the Raman/IR jet spectra of (DCOOD)$_2$ between 1950 and 2450 cm^{-1}. The Raman jet spectra are the same as those in the top trace of Fig. 3.39. The bands marked with * are from DCOOD monomer but not from fundamental modes. See Fig. 3.10 for more details.

Table 3.31: Band positions (in cm^{-1}) and assignments of the Raman active overtone/combination bands of (DCOOD)$_2$ between 1800 and 2450 cm^{-1} in Fig. 3.39, compared with the anharmonic calculation at the MP2/6-311+G(2d,p) level (in cm^{-1}). The experimentally and theoretically derived "anharmonicity constants" of the combination bands ($x_{i,j}$) are given in parentheses. See text for more details.

Label	Experiment Jet[a]	GP[b]	Symmetry	Assignment Jet[a]	GP[b]	Prediction	Calculation
A	1972/1976	1968	A_g	$2\nu_5$	$2\nu_5$	$2 \times 993^a = 1986$	$2000.7\ (-1)$
						$2 \times 994.2^c = 1988$	$1985.4\ (-2)$
B	2051/2062	2042	A_g	$\nu_{20}+\nu_{21}$	$\nu_{20}+\nu_{21}$	$1072.2^c + 994.2^c = 2066$	$2058.3\ (-8)$
C	2080	2081	A_g	$\nu_4+\nu_5$	$\nu_4+\nu_5$	$1093^a + 993^a = 2086\ (-)$	$2116.4\ (-8)$
D, E	2125/2131, 2141	2129	B_g	$\nu_{13}+\nu_{22}$	$\nu_{13}+\nu_{22}$	$890.9^a + 1248.9^a = 2140$	$2138.2\ (-2)$
			A_g	$\nu_6+\nu_{10}$		$1258^a + 894^a = 2152$	$2138.2\ (-2)$
			A_g	$2\nu_{20}$		$2 \times 1072.2^c = 2144$	$2131.4\ (-7)$
F	2160/2166	2165	A_g	$2\nu_4$	$2\nu_4$	$2 \times 1093^a = 2186\ (1)$	$2262.9\ (9)$
G	2187		A_g	$\nu_1?$	ν_2		2167.1
H	2215	2211	A_g	$\nu_{21}+\nu_{22}$		$994.2^c + 1248.9^c = 2243\ (-15)$	$2238.3\ (-5)$
I	2227		A_g	$\nu_5+\nu_6$	$\nu_5+\nu_6$	$993^a + 1258^a = 2251$	$2258.0\ (6)$
J	2247/2252	2239	A_g	$\nu_3+\nu_7$		$1633^a + 624^a = 2257$	$2241.1\ (-6)$
K	2288		B_g	$\nu_{16}+\nu_{18}$		$67.9^c + 2241^d = 2309\ (-8)$	$2304.6\ (-24)$
			A_g	$\nu_3+\nu_{11}$		$67.9^c + 2211^d = 2279\ (9)$	$2304.6\ (14)$
L	2301		B_g	$\nu_{19}+\nu_{22}$		$1633^a + 668^a = 2301\ (0)$	$2304.4\ (0)$
M	2325/2329		B_g	$\nu_{20}+\nu_{22}$		$1072.2^c + 1248.9^c = 2321$	$2321.5\ (-5)$
N	2340/2350/2357/2367/2374	2353	A_g	$\nu_{19}+\nu_{23}$	$\nu_{19}+\nu_{23}$	$1702.7^c + 658.0^c = 2361$	$2355.3\ (-5)$
			B_g	$\nu_{16}+\nu_{17}$		$1093^a + 1258^a = 2351$	$2367.7\ (-6)$
			A_g	$\nu_4+\nu_6$		$140.3^a + 2211^d = 2351$	$2377.2\ (6)$
			B_g	$\nu_{15}+\nu_{18}$		$2215^a + 157^a = 2372$	$2349.1\ (32)$
			A_g	$\nu_1+\nu_9$		$2220^a + 157^a = 2377$	$2393.6\ (9)$
O	2394/2400	2394	A_g	$\nu_2+\nu_9$		$140.3^a + 2241^d = 2381$	$2383.1\ (-18)$
			B_g	$\nu_{15}+\nu_{17}$		$2215^a + 192^a = 2407$	$2390.7\ (39)$
			A_g	$\nu_1+\nu_8$		$2220^a + 192^a = 2412$	$2427.6\ (9)$
P	2423/2432	2429	B_g	$\nu_{14}+\nu_{19}$	$\nu_{14}+\nu_{19}$	$717.9^a + 1702.7^c = 2421$	$2418.4\ (-2)$
			B_g	$\nu_1+\nu_{12}$		$2215^a + 210^a = 2425$	$2416.9\ (38)$
			B_g	$\nu_2+\nu_{12}$		$2220^a + 210^a = 2430$	$2454.3\ (9)$

[a] Raman jet spectrum, this work.   [b] Bertie and Michaelian, Raman GP spectrum (Ref. 9).

[c] Anharmonic calculation results at the MP2/6-311+G(2d,p) level.   [d] IR jet spectrum, courtesy of Zielke.

Label	Experiment		Assignment	Prediction	Calculation
	Jet[a]	GP[b]			
a	1975	1981	$\nu_5+\nu_{21}$ (B_u)	$993^c+994.2^d=1987$ (−12)	1990.6 (−5)
			$\nu_4+\nu_{13}$ (A_u)	$1093^c+890.9^d=1984$ (−9)	2011.7 (−2)
b	2061		$\nu_5+\nu_{20}$ (B_u)	$993^c+1072.2^d=2065$ (−4)	2069.8 (−4)
c	2075	2073	$\nu_4+\nu_{21}$ (B_u)	$1093^c+994.2^d=2087$ (−12)	2108.0 (−9)
d	2140/2148	2153	$\nu_6+\nu_{13}$ (A_u)	$1258^c+890.9^d=2149$	2139.1 (−2)
			$\nu_{10}+\nu_{22}$ (A_u)	$894^c+1248.9^d=2143$	2137.3 (−2)
e/f/g	2174		$\nu_4+\nu_{20}$ (B_u)	$1093^c+1072.2^d=2165$ (9)	2157.6 (−37)
i, j	2227, 2241	2226, 2248	$\nu_5+\nu_{22}$ (B_u)	$993^c+1248.9^d=2242$	2254.0 (3)
			$\nu_6+\nu_{21}$ (B_u)	$1258^c+994.2^d=2252$	2236.5 (−8)
j1/k/l/m/n[c]	2261/2272/2279/2289/2295/2306/2311/2320	2293/2323	$\nu_1+\nu_{16}$ (A_u)	$2215^c+67.9^d=2283$	2262.8 (28)
			$\nu_2+\nu_{16}$ (A_u)	$2220^c+67.9^d=2288$	2311.1 (9)
			$\nu_3+\nu_{23}$ (B_u)	$1633^c+658.0^d=2291$	2277.1 (2)
			$\nu_7+\nu_{19}$ (B_u)	$624^c+1702.7^d=2327$	2320.5 (−6)
CO2?	2331/2342/2353/2362/2367	2372	$\nu_6+\nu_{20}$ (B_u)	$1258^c+1072.2^d=2330$	2320.4 (−2)
			$\nu_4+\nu_{22}$ (A_u)	$1093^c+1248.9^d=2342$	2371.7 (0)
			$\nu_3+\nu_{14}$ (A_u)	$1633^c+717.9^d=2351$	2350.1 (15)
			$\nu_1+\nu_{15}$ (A_u)	$2215^c+140.3^d=2355$	2343.1 (36)
			$\nu_2+\nu_{15}$ (A_u)	$2220^c+140.3^d=2360$	2383.5 (9)
			$\nu_9+\nu_{18}$ (B_u)	$157^c+2211^a=2368$	2387.8 (7)
			$\nu_{11}+\nu_{19}$ (A_u)	$668^c+1702.7^d=2371$	2374.6 (−2)
o	2399/2407/2420	2432	$\nu_9+\nu_{17}$ (B_u)	$157^c+2241^a=2398$	2389.5 (−22)
			$\nu_8+\nu_{18}$ (B_u)	$192^c+2211^a=2403$	2422.1 (7)
			$\nu_{12}+\nu_{18}$ (A_u)	$210^c+2211^a=2421$	2448.2 (6)
			$\nu_8+\nu_{17}$ (B_u)	$192^c+2241^a=2433$	2429.2 (−16)

[a] IR jet spectrum, courtesy of Zielke.
[b] Millikan and Pitzer, IR GP spectrum (Ref. 75).
[c] Raman jet spectrum, this work.
[d] Anharmonic calculation results at the MP2/6-311+G(2d,p) level.

Table 3.32: Band positions (in cm^{-1}) and assignment of the IR active combination bands of (DCOOD)$_2$ between 1950 and 2450 cm^{-1} in Fig. 3.40, compared with the former IR GP study [75] as well as the anharmonic calculation at the MP2/6-311+G'(2d,p) level (in cm^{-1}). The symmetries, the experimentally and theoretically derived "anharmonicity constants" of the combination bands ($x_{i,j}$) are given in parentheses.

Experiment	Assignment	Prediction	Calculation
1313	$\nu_6 + \nu_{16}$ (A$_u$)	$1258^a + 67.9^b = 1326 \; (-13)$	$1316.9 \; (-1)$
	$\nu_{11} + \nu_{23}$ (A$_u$)	$668^a + 658.0^b = 1326 \; (-13)$	$1323.8 \; (-52)$
	$\nu_4 + \nu_{24}$ (B$_u$)	$1093^a + 240.5^b = 1334 \; (-21)$	$1357.1 \; (-6)$
	$\nu_7 + \nu_{14}$ (A$_u$)	$624^a + 717.9^b = 1342 \; (-29)$	$1338.5 \; (-3)$
1383	$\nu_{11} + \nu_{14}$ (B$_u$)	$668^a + 717.9^b = 1386 \; (-3)$	$1354.0 \; (-38)$
	$\nu_6 + \nu_{15}$ (A$_u$)	$1258^a + 140.3^b = 1398 \; (-15)$	$1388.8 \; (-2)$
	$\nu_9 + \nu_{22}$ (B$_u$)	$157^a + 1248.9^b = 1406 \; (-23)$	$1398.3 \; (-1)$
1446	$\nu_8 + \nu_{22}$ (B$_u$)	$192^a + 1248.9^b = 1441 \; (5)$	$1431.3 \; (-2)$
	$\nu_{12} + \nu_{22}$ (A$_u$)	$210^a + 1248.9^b = 1459 \; (-13)$	$1459.4 \; (-1)$
1596	$\nu_{10} + \nu_{14}$ (B$_u$)	$894^a + 717.9^b = 1612 \; (-16)$	$1607.5 \; (0)$
	$\nu_7 + \nu_{21}$ (B$_u$)	$624^a + 994.2^b = 1618 \; (-22)$	$1617.2 \; (-1)$
1884	$\nu_9 + \nu_{19}$ (B$_u$)	$157^a + 1702.7^b = 1860 \; (24)$	$1850.7 \; (-2)$
	$\nu_7 + \nu_{22}$ (B$_u$)	$624^a + 1248.9^b = 1873 \; (11)$	$1869.8 \; (-3)$
	$\nu_3 + \nu_{24}$ (B$_u$)	$1633^a + 240.5^b = 1874 \; (10)$	$1865.5 \; (8)$
	$\nu_5 + \nu_{13}$ (A$_u$)	$993^a + 890.9^b = 1884 \; (0)$	$1892.7 \; (0)$
	$\nu_{10} + \nu_{21}$ (A$_u$)	$894^a + 994.2^b = 1888 \; (-4)$	$1883.9 \; (0)$
	$\nu_8 + \nu_{19}$ (B$_u$)	$192^a + 1702.7^b = 1895 \; (-11)$	$1885.1 \; (-2)$

[a] Raman jet spectrum, this work.

[b] Anharmonic calculation results at the MP2/6-311+G(2d,p) level.

Table 3.33: Band positions (in cm^{-1}) and assignment of the IR active combination bands of (DCOOD)$_2$ observed in the former GP spectrum [75] below 1950 cm^{-1}, compared with the anharmonic calculation at the MP2/6-311+G(2d,p) level (in cm^{-1}). The symmetries, the experimentally and theoretically derived "anharmonicity constants" of the combination bands ($x_{i,j}$) are given in parentheses. See text for more details.

were observed in the GP spectrum [75]. The band positions and assignments are listed in Tab. 3.33.

At the end of Chap. 3.2.2 we have mentioned that the overtone bands of the intramolecular O–H/D out-of-plane bending modes have relatively large negative theoretical anharmonicity constants. The same case applies to the O–H/D in-plane bending modes (ν_4, ν_{20}) as well (see Tabs. 3.19, 3.2.3, 3.27 and 3.31). Deuteration of the hydroxyl group changes the character of this motion somehow. Positive theoretical anharmonicity constants are found for the overtone bands of ν_4 of the two O-deuterated FAD: 24 cm^{-1} for (HCOOD)$_2$ (see Tab. 3.27) and 9 cm^{-1} for (DCOOD)$_2$ (see Tab. 3.31).

The O–H/D stretching vibrational mode shows a furthermore interesting character. All the Raman/IR active combination bands with the Raman active O–H/D stretching mode ν_1 as a combination component have relatively large positive theoretical anharmonicity constants (more than $30\,\text{cm}^{-1}$) whereas those with the IR active fundamental ν_{17} all have large negative theoretical anharmonicity constants (more than $-15\,\text{cm}^{-1}$). The theoretical anharmonicity constants of all the combination bands with a C–H/D stretching mode (ν_2/ν_{18}) as a combination component are positive, and small.

3.3 Conclusions

The Raman spectra of formic acid and its three deuterium isotopomers have been measured in the entire fundamental wavenumber region. The fundamental modes of the monomer and dimer have been assigned and compared with former experimental results as well as with quantum chemical calculations. No solid evidence for the existence of the energetically less stable *cis*-FA monomer was found in our spectra.

Raman active fundamentals of the four isotopomers of FAD and the IR active fundamentals of $(HCOOH)_2$ measured in this work are used as a "benchmarking" to test the accuracy of the calculations (see Tabs. 3.7 and 3.13), because thermal shifts in the room temperature gas phase are an order of magnitude larger and matrix isolation shifts are difficult to predict (A probable exception is the carboxyl stretching whose wavenumber is very sensitive to the conformational temperature, the experimental results in an Ar-matrix fit better to the anharmonic calculations due to the lower temperature during the measuerment). The calculation results at the best-fitting method/level is used conversely to test the experimental results of different measurement methods when a jet spectrum is unavailable. For the intermolecular modes, the harmonic calculation at the simple MP2/6-31+G* level fits best to the experimental results [4], even compared to the many anharmonic calculations. For the entire fundamental wavenumber region, the anharmonic calculations at the MP2/6-311+G(2d,p) level show the best performance. The "most" accurate experimental band positions of the fundamental modes of the four isotopomers of FAD are listed in Tab. 3.34, compared with the theoretical predictions.

Raman active overtone/combination bands of FAD in the entire fundamental wavenumber region have been analyzed. There are a total of 24 overtone bands, all of which are Raman-active, and 276 ($24 \times 23/2$) binary combination bands of FAD, half of which (138) are Raman-active. Not all of them could be detected in our Raman jet spectra due to different Raman scattering activities. About 70-80 overtone/combination bands were observed and assigned for each isotopomer, most of which have A_g symmetry. Results were compared with former studies. A small amount of them cannot be successfully assigned by binary combinations. Ternary or even more complicated coupling mechanismus must be postulated.

Several IR active bands are also assigned, mainly in the O/C–H/D region. With the help of IR jet spectra, a preliminary analysis of the bands in the O/C–H/D region is made, which may contribute to the construction of an anharmonic force field of FAD. The bands are believed to gain extra intensity from coupling to the IR and Raman active O–H stretching modes of the appropriate symmetry [88, 102]. Therefore, the spectra of $(DCOOH)_2$ and $(HCOOD)_2$ show a simpler structure and lower band intensities, as C-H/O–H or C–D/O–D overlap is avoided. Raman jet spectroscopy has confirmed its effectiveness as a tool for the investigation of hydrogen bonded clusters. Compared to gas phase Raman spectra, the jet spectra show major simplifications and bandwidth reductions, which make the analysis possible.

The C/O–H/D stretching fundamentals of formic acid monomer and dimer observed in this work are summarized in Tab. 3.35. For the monomer bands, most of the experimental values fit very well with the anharmonic calculations (see Tab. 3.35). The wavenumber shifts are all within $11 \, \mathrm{cm}^{-1}$ except the ν_2 band of HCOOD. The wavenumber difference between the measured and calculated values of this band is $31 \, \mathrm{cm}^{-1}$. However, the relative deviation is also around only 1%. For the dimer C–H/D fundamentals, the calculations show also a very good agreement in the comparison, and the largest difference between the measured

Description	Mode	(HCOOH)2		(DCOOH)2		(HCOOD)2		(DCOOD)2	
		Experiment	Calculation	Experiment	Calculation	Experiment	Calculation	Experiment	Calculation
A_g									
ν(O-H)	ν_1		2914 (59.6)	....	2786 (253.6)	2200[d]	2158 (124.8)	2215[a]	2167 (75.8)
ν(C-H)	ν_2	**2952**[a]	2950 (469.7)	**2212**[a]	2235 (130.1)	**2954**[a]	2971 (270.5)	**2220**[a]	2234 (173.7)
ν(C=O)	ν_3	**1668**[a]	1657 (35.8)	**1643**[a]	1635 (45.1)	**1653**[a]	1636 (47.3)	**1633**[a]	1617 (55.0)
δ(O-H)	ν_4	**1431**[a]	1443 (11.0)	**1418**[a]	1438 (5.7)	1076[a]	1097 (5.2)	1093[a]	1123 (5.2)
δ(C-H)	ν_5	**1376**[a]	1386 (8.1)	1001[a]	1011 (4.4)	**1382**[a]	1394 (9.9)	993[a]	1002 (4.2)
ν(C-O)	ν_6	**1224**[a]	1219 (10.7)	**1238**[a]	1228 (15.0)	**1268**[a]	1264 (7.5)	**1258**[a]	1250 (11.4)
δ(OCO)	ν_7	**682**[a]	680 (6.2)	**676**[a]	674 (6.2)	**628**[a]	629 (5.5)	**624**[a]	624 (5.5)
δ(O-H...O)	ν_8	194[a]	186 (0.2)	193[a]	184 (0.3)	194[a]	186 (0.3)	192[a]	184 (0.3)
δ(O-H...O)	ν_9	161[a]	153 (1.1)	159[a]	153 (3.4)	157[a]	151 (1.0)	157[a]	150 (1.0)
B_g									
γ(C-H)	ν_{10}	1055/1062[a]	1056 (1.9)	891/897[a]	891 (0.2)	1053/1060[a]	1052 (1.9)	891/897[a]	890 (0.2)
γ(O-H)	ν_{11}	907/915[a]	911 (0.5)	918/924[a]	918 (0.5)	671/677[a]	676 (0.3)	665/670[a]	674 (0.2)
γ(O-H...O)	ν_{12}	239/246[a]	244 (5.2)	210/215[a]	214 (3.4)	235/240[a]	241 (5.0)	207/212[a]	211 (3.4)
A_u									
γ(C-H)	ν_{13}	**1069**[b]	1068 (85.8)	**956/985**[c]	963 (216.4)	1037[d]	1056 (3.9)	**890**[d]	891 (1.5)
γ(O-H)	ν_{14}	939[b]	950 (147.1)	**891/892**[c]	888 (8.1)	693[d]	723 (137.4)	678[d], 669[e]	718 (131.3)
γ(O-H...O)	ν_{15}	**168.5**[f]	169 (8.1)	148[g]	145 (5.5)	**158**[h]	162 (6.9)	**135**[i]	140 (4.9)
twist	ν_{16}	69.2[j]	67 (2.1)	....	67 (2.2)	**68**[h]	68 (2.1)	**68**[i]	68 (2.2)
B_u									
ν(O-H)	ν_{17}	3076[c]?	2990 (2243.7)	3078[c]?	2979 (2477.4)	**2264**[j]	2256 (1333.0)	**2241**[j]	2261 (1077.6)
ν(C-H)	ν_{18}	**2940**[j]	2963 (335.8)	**2211**[j]	2221 (124.3)	**2951**[j]	2972 (64.5)	**2211**[j]	2231 (338.9)
ν(C=O)	ν_{19}	**1741**[b], **1729/1730**[c]	1724 (776.8)	1708/1709/1723[c]	1703 (807.9)	1745[d], **1736**[e]	1719 (739.3)	**1717.5**[k]	1703 (770.4)
δ(O-H)	ν_{20}	**1406**[b]	1414 (4.7)	1381/84[c]	1380 (7.6)	1037[d]	1056 (80.3)	1055[d], **1070**[e]	1072 (15.3)
δ(C-H)	ν_{21}	**1373**[b]	1375 (32.5)	1003/1006[c]	1014 (47.7)	**1387**[d]	1400 (28.4)	976/987[d], 984[e]	994 (70.0)
ν(C-O)	ν_{22}	**1230.2**[j]	1230 (364.9)	1244[c]	1237 (273.7)	**1259**[d], **1249**[e]	1261 (253.9)	**1246**[d], **1249**[e]	1249 (209.6)
δ(OCO)	ν_{23}	**712**[m]	707 (38.3)	695[d]	701 (38.6)	651[d], 675[e]	663 (46.6)	642[d]	658 (46.2)
δ(O-H...O)	ν_{24}	264[b]	254 (68.9)	240[g]	249 (65.6)	240[h]	246 (65.7)	227[i]	240 (62.7)

[a] Raman jet spectra, this work.
[b] IR jet spectra, courtesy of Kollipost.
[c] Ito, Ar matrix isolation IR spectrum (Ref. 15).
[d] Millikan and Pitzer, IR GP spectra at room temperature (Ref. 75.)
[e] Maréchal, IR GP spectra at room temperature (Ref. 83).
[f] Georges et al., high resolution FIR IR spectrum (Ref. 11).
[g] Jakobsen et al., FIR spectrum in solid nitrogen matrix (Ref. 106).
[h] Carlson et al., FIR GP spectrum (Ref. 105).
[i] Clague and Novak, FIR GP spectrum (Ref. 113).
[j] Ito, jet-cooled cavity ring-down IR spectrum (Ref. 95).
[k] Gutberlet et al., high resolution FIR GP spectra (Ref. 85).
[l] Ito, jet spectra, courtesy of Zielke.
[m] Halupka and Sauder, Ar matrix isolation IR spectrum (Ref. 13).

Table 3.34: Band positions (in cm^{-1}) and assignment of the 24 fundamental modes of (HCOOH)2 and its isotopomers, compared with the anharmonic calculation at the MP2/6-311+G(2d,p) level (in cm^{-1}). The experimental results with wavenumber shifts within 1% of the calculations are given in bold face. The IR intensities (in km/mol) and Raman activities (in Å^4/u) from the harmonic calculations at the same level are listed in parentheses.

Mode	HCOOH		DCOOH		HCOOD		DCOOD	
	Exp.[a]	Cal.[b]	Exp.[a]	Cal.[b]	Exp.[a]	Cal.[b]	Exp.[a]	Cal.[b]
ν_1^{OH}	3570	3746/3560	...	3746/3561	2632	2724/2623	2633	2724/2622
ν_2^{CH}	2943	3115/2950	2220	2314/2226	2939	3116/2970	2232	2313/2242

Mode	(HCOOH)$_2$		(DCOOH)$_2$		(HCOOD)$_2$		(DCOOD)$_2$	
	Exp.	Cal.[b]	Exp.	Cal.[b]	Exp.	Cal.[b]	Exp.	Cal.[b]
ν_1^{OH}	...	3144/2914	...	3141/2786	2200[c]	2303/2158	2215[c]	2300/2167
ν_2^{CH}	2952[c]	3127/2950	2212[c]	2322/2235	2954[c]	3130/2971	2220[c]	2321/2234
ν_{17}^{OH}	...	3254/2990	...	3250/2979	2236[d]	2369/2256	2241[d]	2373/2261
ν_{18}^{CH}	2940[d]	3124/2963	2211[d]	2319/2221	2951[d]	3129/2972	2211[d]	2315/2231

[a] Raman GP spectra, this work.
[b] Harmonic/anharmonic calculation results at the MP2/6-311+G(2d,p) level.
[c] Raman jet spectra, this work.
[d] IR jet spectra, courtesy of Zielke.

Table 3.35: Comparison of the experimental and calculated wavenumbers (in cm^{-1}) of FAD C/O–H/D stretching fundamental modes.

and calculated wavenumber is $23\,\mathrm{cm}^{-1}$ from ν_{18} of (HCOOH)$_2$, which means all the deviations are less than 1%. For the O–H/D fundamental modes of the dimer, there are relative large wavenumber differences between experiment and theory, and the deviations for the ν_1 mode of the dimer even rise to about 2% ($\sim 50\,\mathrm{cm}^{-1}$).

After forming the two strong intermolecular O–H/D…O bonds, the wavenumbers of the dimer O–H/D stretching fundamentals decrease much ($\geq 390\,\mathrm{cm}^{-1}$), for both IR and Raman active modes. Both harmonic and anharmonic calculations indict that the Raman active mode ν_1 has always a lower wavenumber than that of its IR active Davydov partner ν_{17}. The experimental results of the two O-deuterated isotopomers confirm this prediction.

In contrast, the C–H/D stretching fundamentals of the dimer are not much influenced by the new hydrogen bonds and remain almost at the same wavenumbers as the related monomer bands. For (HCOOH)$_2$, the wavenumber of the monomer band ν_2 is between the Davydov pair (ν_2 and ν_{18}) of the dimer. It does not fit with the calculations, which show that both dimer bands have higher wavenumbers that the monomer one, although the experimental value of ν_{18} of the dimer is only $3\,\mathrm{cm}^{-1}$ lower than the ν_2 band of the monomer. The experiment shows also the Raman active mode ν_2 is $12\,\mathrm{cm}^{-1}$ higher than the IR active one. It is only in agreement with the harmonic calculation (see Tab. 3.35). The experimental values of ν_2 and ν_{18} modes of $(DCOOH)_2$ are similar and lower than that of the C–D stretching fundamental ν_2 of the monomer. The harmonic calculation indicates that the wavenumbers of the Davydov pair are similar, but both are higher than ν_2 band of the monomer, whereas the anharmonic calculation shows the wavenumber of ν_2 band of the monomer is between the two dimer bands. The comparison of the experimental values fits the harmonic calculation completely for the O-deuterated formic acid, whereas the anharmonic calculation shows the right wavenumber sequence of the three bands mentioned above for the fully deuterated isotopomer, not only between the related monomer and dimer but also between the Davydov pairs. The comparisons of the experimental results with the anharmonic calculations are shown in Fig. 3.41.

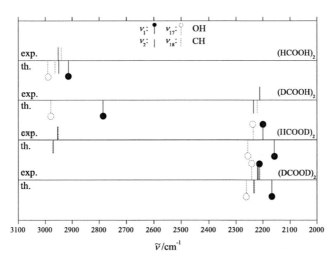

Figure 3.41: Comparison of the predicted and experimental band positions of the FAD C/O–H/D stretching fundamental modes. The Raman active bands are marked with black solid lines while the IR active ones are shown as red dotted lines. See also Tab. 3.35.

The influences of the H/D isotope exchange on the C/O atom on the O/C–H/D stretching modes are analyzed in Tab. 3.36. Generally the wavenumber shifts caused by this isotope exchange are rather small, except that between the calculated Raman active O–H stretching fundamentals (ν_1) of (HCOOH)$_2$ and (DCOOH)$_2$ at the anharmonic MP2/6-311+G(2d,p) level. The ν_1 band of (DCOOH)$_2$ has a much lower wavenumber because this band has a very strong Fermi resonance (reduced cubic constant $= -114.339$) with the combination band $\nu_3 + \nu_4$ during the calculation.

The wavenumber shifts are not always in agreement with the calculations at the MP2/6-311+G(2d,p) level. Nevertheless, some information from the comparison is useful for the band assignments. For example, there is a group of intense bands between $2210\,\text{cm}^{-1}$ and $2264\,\text{cm}^{-1}$ in the IR jet spectrum of (HCOOD)$_2$ (see Fig. 3.37), all of which are possible to be assigned as the O–D stretching fundamental mode ν_{17} ($2256\,\text{cm}^{-1}$ according to the anharmonic calculation at the MP2/6-311+G(2d,p) level). Band j at $2236\,\text{cm}^{-1}$ is finally chosen because it has the best performance in all the comparisons, not only in Tab. 3.36, but also in the wavenumber sequence mentioned above.

What renders the Raman data particularly valuable is the potential for large tunneling splittings in suitable excitation states due to centrosymmetric double proton exchange [114, 121, 122]. In contrast to IR active modes, where excitation of an exchange promoting motion in one hydrogen bond is always coupled with the opposite phase motion in the other hydrogen bond and may thus increase [89, 111] or decrease [79, 93, 107] the splitting relatively slightly, one can hope for much larger tunneling signatures observable even at comparatively low spectral resolution in the Raman spectrum. Like in malonaldehyde, one of the most elementary model systems for single hydrogen transfer in the electronic ground state [5], such experimental benchmark values should prove valuable in the further development of theoretical models for concerted double proton transfer.

The vibrational ground state tunnelling splitting of $0.015\,\text{cm}^{-1}$ was deduced by Havenith and coworkers in a series of high-resolution IR studies [123]. The tunnelling dynamics is dominated by very large zero point effects, which produce

Monomer

C–D→C–H	O–H (ν_1)	O–D (ν_1)
exp.	···· 3570	2633→2632 (−1)
harm. cal.	3746→3746 (0)	2724→2724 (**0**)
anharm. cal.	3561→3560 (−1)	2622→2623 (1)

O–D→O–H	C–H (ν_2)	C–D (ν_2)
exp.	2939→2943 (4)	2232→2220 (−12)
harm. cal.	3116→3115 (**−1**)	2313→2314 (1)
anharm. cal.	2970→2950 (−20)	2242→2226 (**−16**)

Dimer

C–D→C–H	O–H (ν_{17})	O–H (ν_1)	O–D (ν_1)	O–D (ν_{17})
exp.	...	...	2215→2200 (5)	2241→2236 (−5)
harm. cal.	3250→3254 (4)	3141→3144 (3)	2300→2303 (**3**)	2373→2369 (−4)
anharm. cal.	2979→2990 (−11)	2786→2914 (−128)	2167→2158 (−9)	2261→2256 (**−5**)

O–D→O–H	O–H (ν_2)	O–H (ν_{18})	O–D (ν_2)	O–D (ν_{18})
exp.	2954→2952 (−2)	2951→2940 (−11)	2220→2212 (−8)	2211→2211 (0)
harm. cal.	3130→3127 (**−3**)	3129→3124 (−5)	2321→2322 (1)	2315→2319 (**4**)
anharm. cal.	2971→2950 (−21)	2972→2963 (**−9**)	2234→2235 (1)	2231→2221 (−10)

Table 3.36: Analysis of the influence of the H/D isotope exchange on the O/C–H/D stretching fundamentals. The experimental wavenumbers (in cm⁻¹) are from this work and the harmonic/anharmonic wavenumbers (in cm⁻¹) are calculated at the MP2/6-311+G(2d,p) level. The wavenumber shifts are listed in parentheses and those calculation results in better agreement with the experimental results are given in bold face.

variations of tunnelling times of 1-2 orders of magnitude upon vibrational excitation [122]. Even though the small splitting is not observed in our Raman spectra. A largely improved experimental setup will be needed.

4 Acetic acid dimer

In the gas phase acetic acid forms a largely planar cyclic complex with two equivalent hydrogen bonds, the same as in the case of formic acid. Vibrational spectra can provide information about the local properties of this energetically stable cyclic acetic acid dimer (AAD) and due to its centrosymmetry, infrared and Raman techniques are perfectly complementary experimental approaches. However, the progress in applying these two methods has differed. Infrared spectra of AAD fundamentals have been investigated in non-condensed phase for a long time, both in the gas phase (GP) close to room temperature [83,105,113,124–126] and combined with supersonic jet expansion [38,42,101,102]. In contrast, Raman spectroscopic studies in non-condensed phase are rare. The first gas-phase Raman spectra of acetic acid were measured by Gaufrés et al. at 220°C and the dimer bands showed strong thermal broadening [127]. Bertie et al. assigned Raman active bands in the gas phase spectrum at room temperature in 1982 [128]. All the reported Raman spectra are for $(CH_3COOH)_2$ $((HAc)_2)$ and no spectra of the deuterated isotopomers were obtained.

Several anharmonic quantum mechanical studies concentrated on the IR spectra of AAD [17,117,129,130], most of which relied on the jet cooled dimer spectra [17,117,129]. Dreyer has simulated the IR-absorption spectra in the O–H stretching vibrational region using density-functional theory, which fit the experimental results quite well [17]. However, the calculated wavenumbers of most of the sub bands cannot be experimentally verified and further used for the anharmonical analysis because the accurate band positions of most Raman active fundamental modes serving as linear combination partners are unknown. This lack of Raman spectra for cold, isolated AAD decelerated the evaluation of quantum mechanical calculations. Only Nakabayashi *et al.* explained some vibrations of the liquid structure of acetic acid with Raman spectroscopy and ab initio molecule orbital calculations in 1999 [131].

AAD has 42 fundamental modes which are numbered following Herzberg [8], with ν_1 to ν_{14} having A_g symmetry under the C_{2h} point group, ν_{15} to ν_{21} B_g, ν_{22} to ν_{29} A_u, and ν_{30} to ν_{42} B_u. They correspond to 22 different kinds of vibrational motion. While the intermolecular ring twist mode ν_{29} is only IR active and the intermolecular stretching mode ν_{13} is only Raman active, the other 20 vibrations occur in Davydov pairs of IR active and Raman active modes, similar as in the case of FAD. This nomenclature has also been adopted in the later work [128]. The band descriptions used in this work are the same as those in Refs. 132, 133 and may be somehow different from the other work. All the six intermolecular fundamental modes and the methyl torsion mode ν_{29} are in the wavenumber region under $200\,\mathrm{cm}^{-1}$, and all the C/O–H/D stretching modes are higher than $1800\,\mathrm{cm}^{-1}$, spreading over a wide wavenumber range. Therefore, as in the discussion of FAD, the spectra of AAD will be discussed in three wavenumber regions: the low-frequency intermolecular vibration region ($\leq 200\,\mathrm{cm}^{-1}$), the finger-print intramolecular vibration region ($200\text{-}1800\,\mathrm{cm}^{-1}$) and the O–H/D as well as C–H/D vibration region ($\geq 1800\,\mathrm{cm}^{-1}$). The Raman spectra in all these three regions and the IR spectra in the $\geq 1800\,\mathrm{cm}^{-1}$ wavenumber region of $(\mathrm{HAc})_2$ and its singly ($(\mathrm{DAc})_2$) and fully deuterated isotopomers ($(\mathrm{D_4Ac})_2$) will be discussed. The monomer bands will not be assigned.

As already mentioned in Chap. 2.3, the influence of experimental conditions on the band shapes and wavenumbers cannot be neglected in the measurements of AAD. This influence will be discussed in detail in different wavenumber regions.

4.1 The region up to $200\,\mathrm{cm}^{-1}$

Like FAD, AAD has six modes corresponding to the intermonomer vibrations, all under $200\,\mathrm{cm}^{-1}$. Three modes are IR active: ν_{29} (A_u) for the intermolecular twisting vibration, ν_{42} (B_u) for the O–H...O in-plane bending δ(O–H...O) and

ν_{28} (A$_u$) for the O–H...O out-of-plane bending γ(O–H...O). The intermolecular twisting motion is only IR active, whereas the last two modes have corresponding Raman active partner modes: ν_{14} (A$_g$) and ν_{21} (B$_g$). The O–H...O stretching vibration ν(O–H...O) (ν_{13}, (A$_g$)) is only observable in the Raman spectra. Besides, there is a pair of methyl torsion vibration modes (ν_{20}, B$_g$ and ν_{27}, A$_u$) which do not belong to the intermolecular vibrations, but also fall in this wavenumber region.

4.1.1 Raman experimental results

The Raman jet spectra of the three isotopomers of AAD between 50 and 200 cm^{-1} are shown in Fig. 4.1. All the spectra were taken under **Condition I** (see Chap.

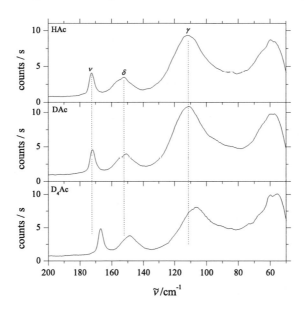

Figure 4.1: Raman jet spectra for three isotopomers of acetic acid between 50 and 200 cm^{-1} under dimer-dominant expansion conditions. See text for details.

145

Description	Mode	(HAc)$_2$			(DAc)$_2$		(D$_4$Ac)$_2$	
		Jet[a]	GP[b]	Liquid[c]	Jet[a]	Liquid[c]	Jet[a]	Liquid[c]
ν(O–H...O)	ν_{13}	173	155	165	172 (1)	165 (0)	167 (6)	160 (5)
δ(O–H...O)	ν_{14}	152	...	55	151 (1)	55 (0)	149 (3)	55 (0)
γ(O–H...O)	ν_{21}	111	98.9	115	111 (0)	112 (3)	106 (5)	108 (7)

[a] Raman jet spectrum, this work.
[b] Bertie and Michaelian, Raman spectrum in gas phase, Ref. 128.
[c] Nielsen and Lund, Raman spectrum in liquid, Ref. 134.

Table 4.1: Band positions (in cm^{-1}) and assignments of the Raman active fundamental modes of (HAc)$_2$ and its OD/D$_4$ isotopomers from Fig. 4.1. The isotope red shifts (in cm^{-1}) relative to the parent bands of (HAc)$_2$ are listed in parentheses.

2.3) to avoid contributions by larger clusters to the band shapes and positions: 0.9% acid in helium ($T_s = 10°$C), $p_s = 200$ mbar, $d = 0.5$ mm, 6×200 s. The measured band maxima are listed in Tab. 4.1, together with previously obtained data in the thermal gas phase [128] and in the liquid [134]. Below 70 cm^{-1}, the onset of the strong methyl torsion band ν_{20} is visible, but its band center cannot be determined reliably due to the Rayleigh edge filter cut-off.

The experimental band positions are compared with several low level quantum chemical approaches in Tab. 4.2. The success of the harmonic MP2/6-31+G* approach in matching experimentally observed anharmonic intermolecular modes of FAD has been emphasized before [4]. Not surprisingly, it also shows the best performance for the low frequency modes of (HAc)$_2$, although the wavenumber

Mode	Exp.	MP2		B3LYP		B97D
		6-31+G*	6-311+G*	6-31+G*	6-311++G(2d,2p)	TZVP
ν_{21} (γ)	111	111.5 (114.7)	92.0	124.0	123.2 (103.8)	120.1
ν_{14} (δ)	152	152.2 (147.2)	152.8	159.3	162.0 (151.6)	166.2
ν_{13} (ν)	173	177.0 (163.4)	168.3	182.1	179.9 (169.7)	176.9

Table 4.2: Comparison of calculated harmonic low frequency fundamentals of (HAc)$_2$ using different quantum chemical methods and basis sets with the anharmonic experimental values (in cm^{-1}). The anharmonic calculation results are listed in parentheses.

Mode	$(HAc)_2$	$(DAc)_2$	$(D_4Ac)_2$
ν_{13} (ν)	177.0	175.9 (1.1)	170.6 (6.4)
ν_{14} (δ)	152.2	149.9 (2.3)	146.8 (5.4)
ν_{20} (τ_{CH_3})	46.3	46.3 (0)	33.8 (12.5)
ν_{21} (γ)	111.5	111.1 (0.4)	103.8 (7.7)

Table 4.3: Calculated harmonic wavenumbers (in cm^{-1}) for Raman active low frequency modes of acetic acid dimer and its isotopomers at the MP2/6-31+G* level. The calculated isotope red shifts (in cm^{-1}) are listed in parentheses.

of the $\nu(O–H\ldots O)$ mode ν_{13} is also more than 2% overestimated like in the case of FAD [4]. The harmonic predictions at this level for the Raman active modes under 200 cm^{-1} of all the three isotopomers are listed in Tab. 4.3.

Several things are notable. The isotope shift of the dimer stretching vibration ν is about 6 to 10 times larger for the fully deuterated acetic acid dimer than for the singly deuterated one. The expected harmonic mass effect is only 3.9. This could in part be due to specific zero point energy effects as in formic acid dimer [3] or it could be due to a different mixing of δ and ν character upon deuteration, considering that the two modes are rather close and have the same A_g symmetry. The latter explanation is supported by harmonic calculations shown in Tab. 4.3. These calculations also predict that the γ mode has the smallest isotopomer shift for $(DAc)_2$ and that the δ mode shifts less in $(D_4Ac)_2$, in full agreement with the experimental assignments. Comparison to thermal gas phase measurements [128] reveals wavenumber discrepancies of more than 10%, mostly due to the weakening of the hydrogen bonds. In the case of the lowest frequency transitions, intensity enhancements in hot band scattering cross sections may reinforce this effect [128]. The δ mode was not assigned in the thermal gas phase. Indeed, as shown in the bottom trace of Fig. 4.2 later, the room temperature gas phase spectrum is not very structured in the ν/δ region.

The Raman spectra of the singly deuterated acid may be compared to the Fourier components obtained from ultrafast vibrational spectroscopy [135], namely 172, 145 and 50 cm^{-1} in CCl_4 solution. The former two clearly correspond to the totally symmetric stretching (ν) and bending (δ) modes which we observe at 172 and $151\,\mathrm{cm}^{-1}$. The out-of-plane γ mode appears to be missing in the ultrafast dynamics, whereas the lowest mode corresponds to the Raman signal we observe at the edge of the spectrum and may be due to methyl torsion. Considering that these modes are superimposed on OD stretching motion in the ultrafast condensed phase experiment, the numerical agreement is excellent but the interpretation is now somewhat different from that available in 2003.

A comparison to liquid state data [134] (see Tab. 4.1) relies on the assumption that the dimer binding motif is conserved in the liquid, despite a chain motif in the crystalline solid [136]. After extensive correction for thermal intensity effects, the stretching mode is indeed in better agreement with the jet-cooled band center than the gas phase value, because the neighboring molecules represent an additional restoring force which compensates for thermal weakening. Isotope effects are described reasonably well. The γ mode is also in good agreement, whereas the in-plane bending mode δ is assigned at far too low wavenumber and clearly involves a misassignment of the methyl torsion due to the proximity of the Rayleigh line, which obscures isotope effects.

4.1.2 Supramolecular stacks of dimers

The effect of thermal excitation by following the spectra of acetic acid dimer through the shock waves limiting the jet expansion are explored. By probing the shock zone between the free jet and the background gas, a continuous evolution of the dimer spectra between cryogenic and ambient temperatures is achieved. By increasing the background pressure p_b in a jet expansion for a given nozzle distance and stagnation pressure p_s, we achieve a rather continuous spectral transition between jet-cooled conditions (upper traces of Fig. 4.2, background pressure < $10\,\mathrm{mbar}$) and conditions in which the dimers are heated during their deceleration in the receding Mach disk zone (background pressure between 20 and $50\,\mathrm{mbar}$). The deceleration is also responsible for part of the signal gain. At $50\,\mathrm{mbar}$ and $1\,\mathrm{mm}$ nozzle distance, the spectrum does not differ much from the conventional

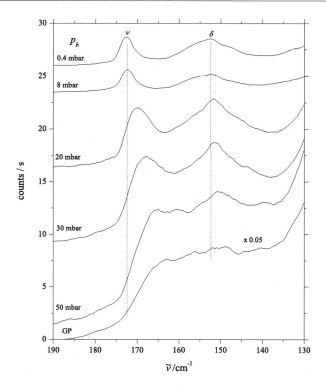

Figure 4.2: Evolution of the Raman jet spectrum of (HAc)$_2$ with increasing background pressure. The jet spectrum on the top was taken under **Condition I**, $6 \times 200\,\mathrm{s}$, and the other jet spectra were taken under the following conditions: $T_\mathrm{s} = 10°\mathrm{C}$ (0.7% HAc in He), $p_\mathrm{s} = 200\,\mathrm{mbar}$, $d = 1\,\mathrm{mm}$, $4 \times 60\,\mathrm{s}$ ($4 \times 200\,\mathrm{s}$ for the spectrum with $p_\mathrm{b} = 8\,\mathrm{mbar}$). Between 8 and 20 mbar, the compression waves start to interfere with the laser zone, resulting in an intensity pile-up due to molecular slow-down and warming of the dimers. A pseudo 300 K gas phase spectrum ($T_\mathrm{s} = 10°\mathrm{C}$, $p_\mathrm{s} = 200\,\mathrm{mbar}$, $p_\mathrm{b} = 80\,\mathrm{mbar}$, $4 \times 30\,\mathrm{s}$) is shown for comparison at the bottom.

gas phase spectrum, qualitatively in line with expectations [137] for such a weak expansion with $\sqrt{p_\mathrm{s}/p_\mathrm{b}} = 2$ and the $4 \times 0.15\,\mathrm{mm}^2$ nozzle. While the predicted Mach disk position [137] is close to the excitation laser for $p_\mathrm{b} = 50\,\mathrm{mbar}$, heating effects can be expected for somewhat smaller background pressures as well. One can thus see how the δ peak emerges from the broad plateau upon cooling of the gas phase (or rather disappears upon shock-heating of the jet expansion) whereas the ν peak arises from (or disappears into) the step region. Both bands shift to lower wavenumber with increasing temperature, in line with the thermal weakening of the hydrogen bonds.

In the context of a comparison to the liquid state spectra, it is also instructive to discuss jet spectra obtained at higher concentration of acetic acid in the carrier gas (saturation pressure at 20°C), where the dimers themselves aggregate further due to dispersion forces, on the way to aerosols [55]. This has two major effects, which may be seen in Fig. 4.3. Due to progressive condensation, the cluster temperature increases and the resulting hot transitions have stronger Raman scattering intensity. This is the effect which was corrected for in the liquid state spectra [134] and it leads to a notable increase of the low frequency slope. A more important effect results from the mutual interaction of the dimers. As Fig. 4.3 shows, only at the shortest nozzle distance (0.5 mm) and under the mildest expansion conditions (0.2 bar stagnation pressure) the spectrum resembles that of dilute expansions (second and third trace from bottom, saturation pressure at 10°C, 0.1-0.2 bar stagnation pressure). With increasing clustering extent, the δ mode and in particular the γ mode shift to higher wavenumber. This can only be explained by the perturbation of these bending vibrations by neighboring dimers. The stronger perturbation of the γ mode indicates a preference for stacking of the planar carboxylic acid dimer units. The dimer stretching mode ν remains almost unaffected. Only at the highest stagnation pressure, a broadening hints at nearest neighbor interactions. These effects suggest that the δ and ν bands merge into a single broad structure in liquid acetic acid, whereas the stronger γ band remains separate, thus explaining the partial misassignment of the liquid state spectra [134]. Remarkably, the effects are much less pronounced in the formic acid dimer case [4]. A possible explanation involves the small size and planarity of that system, which allows for vibrational motion without too much interference with neighboring molecules. We finally note that aggregation beyond dimers can transfer some Raman intensity to IR active dimer modes due to the removal of inversion symmetry. However, this effect is better studied in a less congested higher wavenumber region.

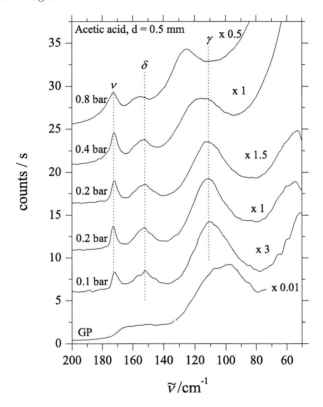

Figure 4.3: Comparison of Raman jet spectra of (HAc)$_2$ at various stagnation pressures, scaled to similar scattering intensity. $d = 0.5\,\text{mm}$, $6 \times 200\,\text{s}$. The upper three traces were recorded at a higher acetic acid concentration (1.6%, $T_s = 20°C$) than the next two (0.7%, $T_s = 10°C$) and show signs of supra-dimeric aggregates. A 300 K gas phase spectrum (the same spectrum as that in Fig. 4.2) is shown for comparison at the bottom.

4.1.3 Former IR studies

It is worthwhile to briefly recapitulate the state of knowledge on the IR active intermolecular modes of acetic acid dimer. They were reported in early low resolution gas phase work [105, 113] and reinvestigated more recently [126]. For the band system detected around 40-60 cm^{-1}, several A_u vibrational assignments are conceivable. These include the hydrogen bond twisting mode ν_{29} ($\tau_{O-H\cdots O}$), the methyl torsion mode ν_{27} (τ_{CH_3}) and possibly the out-of-plane bending (γ) mode ν_{28}, all predicted in this spectral range. The most recent investigation [126] favors the $\tau_{O-H\cdots O}$ and γ modes. A jet-FTIR investigation is currently out of reach [11]. The highest-frequency IR intermolecular mode corresponds to in-plane bending ν_{41} (δ, B_u) and is expected around 170 cm^{-1} (see also Tab. 4.4). Instead of a single band, a pattern with at least two components between 140 and 200 cm^{-1} was observed. Band maxima near 168 and 190 cm^{-1} were found [105, 113]. The most recent explanation involves a Fermi resonance of the B_u band with up to three different combination bands [126]. One of them (with a zero order band position at 138 cm^{-1}) could not be specified, the other two were explained as combinations of $\gamma(A_u)$ with Raman active bands. One of these explanations is still possible in the light of our new results. It is a combination with $\gamma(B_g)$, which should fall within the observed absorption feature, even considering thermal and anharmonicity effects. Such out-of-plane combinations can also have substantial intrinsic intensity, as shown for formic acid dimer [4]. The other proposal, a combination with $\delta(A_g)$, can be safely ruled out as a resonance partner. Firstly, it has the wrong symmetry for a Fermi resonance. Secondly, we now know that $\delta(A_g)$ falls at 152 cm^{-1} and not below 120 cm^{-1}, as conjectured in earlier work [126]. A definitive assignment has to await infrared studies in supersonic jets, but one should not dismiss either the possibility of a single Fermi resonance between $\delta(B_u)$ and $\gamma(A_u) + \gamma(B_g)$ or even no Fermi resonance at all but a simple coexistence of $\delta(B_u)$ around 170 cm^{-1} and $\gamma(A_u) + \delta(A_g)$ around 190 cm^{-1}. A weak combination band observed in the gas phase [105] near 310 cm^{-1} was interpreted differently [72, 128] and may now be tentatively assigned to $\delta(B_u) + \delta(A_g)$.

Description	Mode	Exp.	MP2		B3LYP	
			6-31+G*	6-311+G*	6-31+G*	6-311++G(2d,2p)
τ_{CH_3}	ν_{27}	41	41.4	28.3	44.6	45.0
γ(O–H...O)	ν_{28}	56	70.8	59.7	79.5	77.6
$\tau_{O-H...O}$	ν_{29}	48.5	46.4	34.5	68.7	70.4
δ(O–H...O)	ν_{41}	170.5	170.4	162.4	176.0	178.4

Table 4.4: Comparison of calculated harmonic IR active (HAc)$_2$ low frequency fundamentals using different quantum chemical methods and basis sets with experimental values from Ref. 126 (in cm^{-1}).

A comparison of the acetic acid dimer findings with those for formic acid dimer [4] is instructive. The γ mode drops by more than a factor of two in frequency, because its character partially changes from a tilt of the inner hydrogens to a wagging of the outer substituents, which are much more heavy in acetic acid. For the in-plane bending mode, the wavenumber decrease is less pronounced, indicating that the vibration is still dominated by motion within the carboxylic dimer group. The dimer stretching mode is expected to decrease by a factor of $\sqrt{23/30} = 0.88$, based on simple harmonic mass arguments. The actual factor is 0.91, indicating that the hydrogen bonds in acetic acid dimer are slightly stiffer than in formic acid dimer. We will come back to this observation in Chap. 5.1.2. A remarkable difference is the complete absence of intermolecular combination bands and overtones in the acetic acid dimer spectrum. Despite an intense search up to high concentrations, no transitions equivalent to those observed in particular for out-of-plane motion in formic acid dimer were found [4]. This must reflect the smaller amplitude motion in the heavier frame.

4.2 200-1800 cm^{-1} region

As a continuation and extension of the study on the intermolecular fundamental region, the 13 Raman active fundamentals (ν_4-ν_{12}, ν_{16}-ν_{19}) in the 200-1800 cm^{-1} wavenumber region are assigned and analyzed. With the help of the quantum mechanics calculations and isotope substitution comparison, all the remaining Raman active fundamentals could be assigned except for the very weak out-of-plane O–

H bending mode ν_{18} of the two deuterated isotopomers with an extremely small Raman activity, whose related band of FAD has also a rather low intensity. Comparison of Raman results with the IR active fundamental modes from previous experimental work as well as with some quantum chemical calculations is made. A few combination bands are tentatively assigned. Besides, the red/blue wavenumber shifts of the fundamental modes caused by further aggregation of the dimer are discussed. The remaining four Raman active fundamental modes belong to the O–H/D (ν_1) and C–H/D (ν_2, ν_3 and ν_{15}) stretching vibrations above $1800\,\mathrm{cm}^{-1}$ and will be discussed in the next section.

4.2.1 Fundamental modes of (HAc)$_2$

Raman jet spectra of (HAc)$_2$ were measured under two different conditions and shown in Fig. 4.4. Further on, jet spectra of (DAc)$_2$ and (D$_4$Ac)$_2$ measured only under **Condition I** are shown in Fig. 4.5. The Raman GP spectra of the corresponding compounds are also shown for comparison. All the spectra are combinations of three measurements in different wavenumber regions under equivalent conditions. In the jet spectra, the C–C stretching bands ν_{10} are ten times reduced and some bands are ten times enlarged to fit other bands for a better view. All the monomer bands are marked with M in the spectra. They can be identified by comparison of the jet spectra under different AAD concentrations as well as with the GP spectra. Several bands which do neither belong to the Raman active fundamentals of dimers nor of monomers are marked with different labels in the figures and will be discussed later. A simulated spectrum using the Raman activities calculated at the harmonic B3LYP/6-311++G(2d,2p) level and the anharmonic wavenumbers at the same level is shown in Fig. 4.4 (trace c). The vibrational temperature was set to $100\,\mathrm{K}$ and a full width at half maximum of $8\,\mathrm{cm}^{-1}$ was used for the simulation. It fits to the experimental spectrum quite well.

The measured band maxima and assignment of the fundamental modes of the three isotopomers are summarized in Tab. 4.5. The experimental results of the

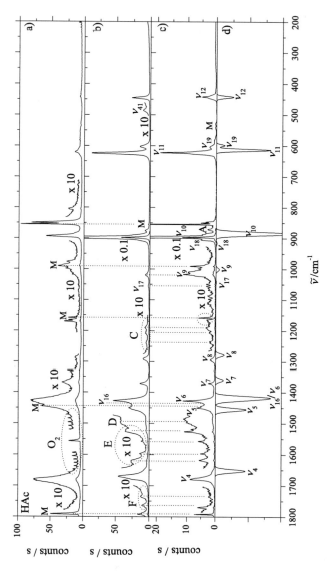

Figure 4.4: Raman spectra of (HAc)$_2$ between 200 and 1800 cm^{-1}: a) GP spectra, $T_s = 20°$C (1.6% HAc in He), $p_b = 30$ mbar, 4×60 s; b) jet spectra under **Condition II**, 12×300 s except the spectrum between 700 and 1250 cm^{-1} (12×120 s) due to the intense C–C stretching band; c) jet spectra under **Condition I**, 6×200 s for the spectrum between 200 and 700 cm^{-1} and 8×300 s for the others; d) simulated spectrum from calculations at the B3LYP/6-311++G(2d,2p) level.

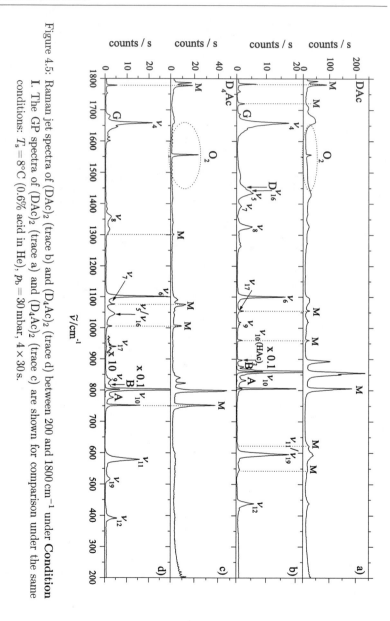

Figure 4.5: Raman jet spectra of $(DAc)_2$ (trace b) and $(D_4Ac)_2$ (trace d) between 200 and 1800 cm^{-1} under **Condition I**. The GP spectra of $(DAc)_2$ (trace a) and $(D_4Ac)_2$ (trace c) are shown for comparison under the same conditions: $T_s = 8°C$ (0.6% acid in He), $p_b = 30$ mbar, 4×30 s.

		Raman active						IR active					
			(HAc)$_2$		(DAc)$_2$	(D$_4$Ac)$_2$		(HAc)$_2$		(DAc)$_2$		(D$_4$Ac)$_2$	
Displacement[a]	Mode	Jet[b]	Jet[c]	GP[d]	Jet[c]	Jet[c]	Mode	S.P.O.[e]	Jet[f]	S.P.O.[e]	Jet[f]	S.P.O.[e]	Jet[f]
	A$_g$						B$_u$						
ν(C=O)	ν_4	1670	1680	1681.5	1659	1659	ν_{33}	1712	1730.8 1733.2	1704 1720	1735 1736.5	1705	1725.9
$\delta_s{}'$(CH$_3$)	ν_5	1456	1450	1428.3	1433	1044	ν_{34}	1431	1439.7	1423	1433.5	1035	…
δ(O–H)	ν_6	1427	1427	1428.3	1100	1101	ν_{35}	1417	1426.2	1072	…	1087	…
δ_s(CH$_3$)	ν_7	1371	1371	1370.2	1392	1080	ν_{36}	1359	1363	1390	1401	1055	…
ν(C–O)	ν_8	1291	1294	1285	1326	1359	ν_{37}	1300	1306.1 1308.3	1326	1333.4 1335.6	1359	1370
ρ_s(CH$_3$)	ν_9	1019	1017	1007	1011	832	ν_{38}	1020	…	1007	…	832	…
ν(C–C)	ν_{10}	898	897	891.7	860	802	ν_{39}	896	…	860	…	810	…
δ(OCO)	ν_{11}	624	622	616.2	595	579	ν_{40}	632	…	613	…	590	…
δ(CCO)	ν_{12}	446	442	438.7	438	391	ν_{41}	484	…	473	…	433	…
	B$_g$						A$_u$						
δ_a(CH$_3$)	ν_{16}	1437	…	…	1440	1044	ν_{23}	1431	1440	…	1433.5	…	…
ρ_a(CH$_3$)	ν_{17}	1056	1055	~1065	1055	940	ν_{24}	1050	…	…	…	…	…
γ(O–H)	ν_{18}	…	906	…	…	…	ν_{25}	956	…	…	…	…	…
γ(CCO)	ν_{19}	603	604	616.2	595	516	ν_{26}	595	…	…	…	…	…

[a] Band notation from Ref. 132. ν means stretching, δ valence angle bending, ρ rocking, γ out of plane bending and τ torsion.
[b] Raman jet spectra under **Condition II**, this work.
[c] Raman jet spectra under **Condition I**, this work.
[d] Bertie and Michaelian, Raman spectrum in gas phase, Ref. 128.
[e] Ovaska, stretched-polymer orientation IR spectrum, Ref. 133.
[f] Häber, FTIR jet spectrum, Ref. 38.

Table 4.5: Summary of the band positions (in cm^{-1}) and assignment of the fundamental modes of (HAc)$_2$ and its OD/D$_4$ isotopomers in the wavenumber region of 200-1800 cm^{-1} of this work and several former studies. See text for more details.

157

IR active fundamentals from M. Ovaska *et al.* [133] using the stretched-polymer orientation method and several bands observed in the FTIR expansion measurements [38, 101] are also listed in Tab. 4.5. Not surprisingly, most of them are blue-shifted relative to the IR data measured at room temperature [133] but the wavenumber differences are mainly within $10\,\mathrm{cm}^{-1}$. The intermonomer coupling of the intramolecular modes in the dimer is rather small: for most Davydov pairs, the differences between the (under jet conditions measured) band positions are less than $20\,\mathrm{cm}^{-1}$. Two exceptions are only the $\nu(\mathrm{C{=}O})$ and the $\gamma(\mathrm{O{-}H})$ modes.

The influence of measurement conditions on the intermolecular bands of AAD has been discussed [54]. The evolution of the spectra with increasing cluster size reveals the stacking structure of supra-dimeric aggregates [54]. The wavenumber differences of the Raman active fundamentals of $(\mathrm{HAc})_2$ observed under two extremely different experimental conditions (see Fig. 4.4) can be found in Tab. 4.5 and the influences on the intramolecular modes can be clearly seen. Most bands are at higher wavenumbers under **Condition II**, due to the mutual interaction of the neighboring dimers. A similar effect was observed in some previous IR experiments [55]. Under rapid collisional cooling the acetic acid molecules can aggregate either into amorphous aggregates or into small crystals with a planar chain structure [55]. With increasing clustering extent, most of the bands shift to higher wavenumbers, compared with the GP as well as jet measurement with the dimer-dominant conditions.

Exceptions are the $\mathrm{C{=}O}$ (ν_4) and $\mathrm{C{-}O}$ stretching modes (ν_8), which have obviously lower wavenumbers under **Condition II**. The red shift of ν_4 can even amount to $10\,\mathrm{cm}^{-1}$ (see also Fig. 4.8 later on). Therefore, larger aggregates of AAD are supposed to form not only chain structures but also double-layer or "sandwich-like" structures, which can be built by two or more parallel layers of dimer (see Fig. 4.6). The $\mathrm{C{=}O}/\mathrm{C{-}O}$ bonds are weakened by the interactions of the hydrogen atoms on the other layer. The lower force constants decrease directly the wavenumbers of the related stretching vibrations and maybe also affects the OCO out-of-plane bending mode ν_{19}, which shows a slight red shift of $1\,\mathrm{cm}^{-1}$.

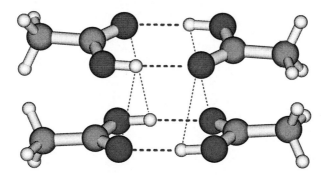

Figure 4.6: Double-layer-structure of the AAD aggregates. The additional interactions of the oxygen atoms with the hydrogen atoms from the neighboring dimer are marked with finer dots.

It is important to note that aggregation beyond dimers can transfer some Raman intensity to IR active dimer modes due to the removal of inversion symmetry. The formed sandwich-like dimer structure has already no more strict C_{2h} symmetry and the IR-Raman exclusion rule does not apply. For example, the weak band at 486 cm⁻¹ observed in the Raman spectra of $(HAc)_2$ (see Fig. 4.4, trace a) and the related band of $(DAc)_2$ at 476 cm⁻¹ as well as $(D_4Ac)_2$ at 433 cm⁻¹ correspond to the IR active OCO in-plane bending mode ν_{41}. The band positions fit very well to the previous IR measurements [133] (see Tab. 4.5) as well as the calculations. Another evidence is shown in Fig. 4.7: its intensity increases with longer nozzle distances, which equal lower species concentration and lower vibrational temperature. Only a band due to larger clusters can rise under these conditions.

The fundamental modes of $(HAc)_2$ are discussed at first, from low to high frequency. Most of them have already been assigned in the previous Raman GP work [128] (see Tab. 4.5) and there is no conflict between that and our assignment. Most of the wavenumber differences are below 10 cm⁻¹.

Due to their band width, several closely spaced bands could not be distinguished in the GP spectrum at room temperature [128] and require further discussion: the OCO in-plane bending mode ν_{11} and the CCO out-of-plane bending mode ν_{19}

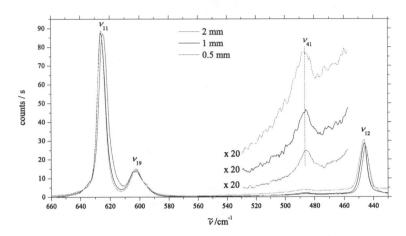

Figure 4.7: Comparison of the band intensities of the IR active OCO in-plane bending mode ν_{41} in the Raman jet spectra under the same measurement conditions but with different nozzle distances: $T_s = 20°C$ (1.6% HAc in He), $p_s = 700\,\text{mbar}$, $12 \times 300\,\text{s}$. The Raman active fundamentals are scaled to similar scattering intensity.

were assigned at $616.2\,\text{cm}^{-1}$ together, whereas symmetric methyl in-plane bending mode ν_5 and the O–H in-plane bending mode ν_6 were mixed at $1428.3\,\text{cm}^{-1}$. Besides, two modes with B_g symmetry (asymmetric methyl in-plane bending mode ν_{16} and the O–H out-of-plane bending mode ν_{18}) were not reported.

Background information from the quantum mechanics calculations can assist the discussion. Tab. 4.6 lists harmonic predictions for the Raman active fundamentals of $(\text{HAc})_2$ at different levels using a variety of basis sets. For the case of ν_{11} and ν_{19}, two related bands were observed in our jet spectra due to the relatively narrow band widths: one band at $604\,\text{cm}^{-1}$ and another one at $622\,\text{cm}^{-1}$. The latter one is very intense and could explain why the former one could not be verified from the GP measurement. The calculations show that ν_{11} has a wavenumber of about $20\,\text{cm}^{-1}$ higher than that of ν_{19} and the assignment is therefore straightforward.

Mode[a]	Exp.	MP2		B3LYP		
		6-31+G*	6-311+G*	6-31+G*	6-311+G*	6-311++G(2d,2p)
ν_4	1680	1751.1	1759.2	1726.6	1724.0	1702.4 (1652.9)
ν_5	1450	1524.1	1511.9	1496.2	1490.1	1503.9 (**1455.5**)
ν_6	1427	1485.4	1477.9	1474.7	1460.4	1470.3 (**1416.6**)
ν_7	1371	1321.7	1424.7	1416.5	1407.9	1405.6 (**1361.4**)
ν_8	1294	1354.8	1317.9	1319.2	1307.4	1317.4 (1278.7)
ν_9	1017	1056.5	1049.5	1036.7	1033.1	1031.5 (1001.4)
ν_{10}	897	917.5	915.5	**903.5**	**899.8**	**901.8** (**885.1**)
ν_{11}	622	**619.9**	**626.9**	**621.6**	**623.9**	**626.7** (**613.9**)
ν_{12}	442	**444.7**	**445.9**	**440.3**	**440.4**	**444.7** (**443.1**)
ν_{16}	1437[b]	1520.9	1509.9	1499.8	1493.3	1483.2 (**1424.5**)
ν_{17}	1055	1089.7	1081.0	1080.8	1078.1	1075.6 (**1043.2**)
ν_{18}	906	937.7	786.1	966.7	928.0	996.2 (**916.3**)
ν_{19}	604	588.2	575.7	**606.2**	**604.0**	611.0 (**598.9**)

[a] See Tab. 4.5 for details of the band descriptions.
[b] Measured under **Condition II**, see text for details.

Table 4.6: Comparison of calculated harmonic fundamentals of $(HAc)_2$ in the wavenumber region of 200-1800 cm^{-1} using different quantum chemical methods and basis sets with jet experimental (anharmonic) values (in cm^{-1}) taken under **Condition I**. The results of the anharmonic calculation at the B3LYP/6-311++G(2d,2p) level are listed in parentheses, whereas the calculations with wavenumber shifts within 13 cm^{-1} of the related experimental values are given in bold face.

For the case of ν_5 (A_g) and ν_6 (A_g), calculations at all levels show that the wavenumber of ν_6 is obviously lower than ν_5. However, most calculations also point out that ν_{16} (B_g) not observed in the GP spectrum is very close to ν_5 (see Tab. 4.6). Two strong bands at 1450 and 1427 cm^{-1} as well as a weak band at 1445 cm^{-1} on the shoulder of the higher wavenumber band were detected under **Condition I**. There is no doubt that the band at 1427 cm^{-1} belongs to ν_6. Depolarization measurements were made to determine the symmetry of the remaining two bands. Trace a) in Fig. 4.8 was recorded with perpendicular polarization of the excitation laser under **Condition I** and trace b) shows the polarized component alone. The latter is estimated by subtracting the spectrum with parallel laser polarization, multiplied by 7/6, from the spectrum with perpendicular laser polarization. All the three mentioned bands as well as a new band at $1435 \, \text{cm}^{-1}$ coming from the spectra subtraction are with A_g symmetry. It should be noted that some of them may come from monomer, as there are also monomer bands with similar wavenumbers (see Fig. 4.4, GP spectrum, trace a)).

Therefore, a depolarization experiment was carried out under **Condition II** to avoid the interference of monomer (trace c) and d) in Fig. 4.8). It is seen that both the bands at $1435 \, \text{cm}^{-1}$ and $1445 \, \text{cm}^{-1}$ disappear. The strong band at $1450 \, \text{cm}^{-1}$ under **Condition I** shifts to $1456 \, \text{cm}^{-1}$ and remains due to its A_g symmetry. A new band at $1437 \, \text{cm}^{-1}$ has B_g symmetry and is assigned to ν_{16}. The only remaining band at $1450 \, \text{cm}^{-1}$ is therefore ν_5. The two dimer bands ν_6, ν_{16} and the monomer band between them are so close that the latter two appear only as an unclear band shoulder of ν_6 in the Raman jet spectrum under **Condition I** (trace a), Fig. 4.8) due to the band overlap. Under **Condition II** (trace c), Fig. 4.8), the monomer band in the middle disappears and the remaining two bands are sufficiently far away to be distinguished. Besides, the different shift directions of the bands in this wavenumber region can be clearly observed. With the influence of larger cluster aggregates, the carboxyl stretching mode ν_4 shifts strongly whereas the C–O stretching fundamental ν_8 shifts slightly to lower wavenumber. ν_5 is strongly blue shifted ($6 \, \text{cm}^{-1}$) but ν_7 very little. ν_6 remains almost unaffected, whereas ν_{16} appears to blue-shift somewhat. ν_8 disappears completely in traces b and d because it has a depolarization ratio $\rho_\perp$ between 0.72 and 0.74, according to the calculations. This band is almost depolarized despite its A_g symmetry.

Finally, because the assignment of the Raman-active fundamentals of $(\text{HAc})_2$ is beyond doubt, it is instructive to compare the experimental anharmonic transitions with harmonic predictions at different levels of quantum-chemical approximation in Tab. 4.6. Most of the calculated wavenumbers fit satisfactorily to the experimental results. It is obvious that the B3LYP level shows better agreement than MP2, but also with a 2-3% systematical overestimation for the modes above $1000 \, \text{cm}^{-1}$. MP2/6-31+G* fits best to the experimental values of the intermolecular modes of the cyclic carboxylic acid dimers [4,54], but not for the intramolecular modes any more. From all levels of harmonic calculation carried out, with and

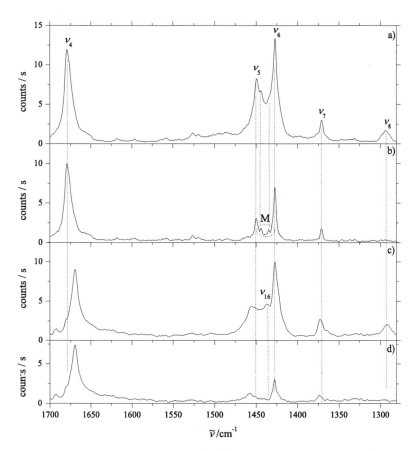

Figure 4.8: Depolarization analysis of (HAc)$_2$ between 1280 and 1700 cm^{-1} under different measurement conditions: a) and b) were under **Condition I** $(6 \times 300\,\text{s})$ whereas c) and d) under **Condition II** $(12 \times 300\,\text{s})$. The top trace of each group is with the excitation laser perpendicular to the scattering plane and the bottom one is the residual after subtracting 7/6 of the spectrum obtained with the excitation laser parallel to the scattering plane.

without diffuse function, the wavenumbers do not vary much with the size of the basis set whereas the difference between the harmonic and anharmonic calculations is rather large. In the case of FAD, the anharmonic calculations at the MP2/6-311+G(2d,p) level fit best to the jet experimental results of the intramolecular fundamentals. For AAD, calculations at the same level were not carried out due to the high calculation costs. Another reason is that the anharmonic calculation at the B3LYP/6-311++G(2d,2p) level shows already a very good agreement for most experimental values, especially for the IR active fundamentals (see Tabs. 4.5 and 4.7), although an underestimation by up to 2% exists for several bands (ν_4, ν_{16}), in contrast to the harmonic wavenumbers. The anharmonic calculation also gives the best answer for the three nearby bands: ν_5, ν_{16} and ν_6. Therefore, the calculation results at this level are used for the spectrum simulation (trace c) of Fig. 4.4). Comparisons of the calculated fundamentals of the three AAD isotopomers wavenumbers at harmonic/anharmonic B3LYP/6-311++G(2d,2p) level are listed in Tab. 4.7 to assist the further assignments.

4.2.2 Fundamental modes of $(DAc)_2$ and $(D_4Ac)_2$

Assisted by the calculations listed in Tab. 4.7, the assignments of the Raman active fundamentals of $(DAc)_2$ and $(D_4Ac)_2$ in Tab. 4.5 are straightforward. The calculated wavenumber shifts between AAD and its isotopically substituted counterparts in the present work match the experimental results very well too.

Several things are notable. It is logical that the two vibrational modes of the O–H/D group, ν_6 and ν_{18} are more sensitive to O–H than C–H deuteration. As expected, similar character was observed for the OCO in-plane bending mode ν_{11}, like for the related mode of FAD (see Tab. 3.11). Not surprisingly, modes corresponding to the various movements of the methyl group (ν_5, ν_7, ν_9, ν_{16} and ν_{17}) are largely sensitive to its deuteration. The two CCO bending modes ν_{19} and ν_{12} show the same isotope dependence as methyl motion modes mentioned above because the CCO chain includes the carbon atom of the methyl group. Wavenumbers of the C–O stretching vibration ν_8 and the C=O stretching mode

Mode	(HAc)$_2$	(DAc)$_2$	(D$_4$Ac)$_2$	Mode	(HAc)$_2$	(DAc)$_2$	(D$_4$Ac)$_2$
A$_g$				**B$_u$**			
ν_4	1702.4/1652.9	1668.8/1615.6	1663.0/1612.6	ν_{33}	1752.4/1704.7	1742.1/1709.9	1735.8/1698.0
ν_5	1503.9/1455.5	1474.8/1430.2	1055.9/1043.9	ν_{34}	1479.7/1425.8	1474.4/1429.6	1055.4/1041.7
ν_6	1470.3/1416.6	1132.7/1101.5	1130.2/1100.6	ν_{35}	1467.6/1412.0	1102.3/1096.0	1114.3/1079.2
ν_7	1405.6/1361.4	1419.0/1371.9	1104.4/1075.1	ν_{36}	1403.6/1358.6	1424.8/1380.9	1087.1/1070.9
ν_8	1317.4/1278.7	1354.8/1320.0	1371.4/1323.3	ν_{37}	1335.1/1297.8	1359.9/1328.5	1382.9/1340.2
ν_9	1031.5/1001.4	1026.8/1003.4	840.8/829.4	ν_{38}	1035.1/1006.6	1027.9/1003.7	845.0/832.2
ν_{10}	901.8/885.1	866.4/848.7	809.2/794.4	ν_{39}	905.5/888.3	871.4/852.9	822.3/812.9
ν_{11}	626.7/613.9	601.1/588.9	583.9/577.1	ν_{40}	637.2/620.4	622.5/606.7	597.2/592.1
ν_{12}	444.7/443.1	439.7/435.9	394.3/391.2	ν_{41}	484.8/480.9	476.4/469.9	434.9/429.1
B$_g$				**A$_u$**			
ν_{16}	1483.2/1424.5	1483.1/1432.4	1068.5/1043.8	ν_{23}	1483.3/1424.6	1483.1/1439.0	1068.5/1043.8
ν_{17}	1075.6/1043.2	1075.4/1051.9	940.1/927.8	ν_{24}	1076.3/1044.2	1076.3/1051.6	942.6/928.2
ν_{18}	996.2/916.3	722.7/682.8	720.2/677.9	ν_{25}	1033.8/961.3	765.8/728.9	756.6/717.8
ν_{19}	611.0/598.9	606.0/589.9	524.8/512.8	ν_{26}	604.0/591.0	588.7/570.2	510.9/498.3

Table 4.7: Comparison of harmonic/anharmonic calculations of the intramolecular fundamentals of (HAc)$_2$ and its OD/D$_4$ isotopomers in the wavenumber region of 200-1800 cm^{-1} at the B3LYP/6-311++G(2d,2p) level (in cm^{-1}).

ν_4 of the three species do not vary much. All the IR active fundamentals behave similar to their related Raman active Davydov partners.

ν_{19} and ν_{11} of $(DAc)_2$ cannot be resolved in the spectra because they are very close to each other according to the anharmonic calculations ($589.9\,cm^{-1}$ for ν_{19} and $588.9\,cm^{-1}$ for ν_{11}, see Tab. 4.7). The same is true for ν_5 and ν_{16}. The experimental wavenumber difference of them is $13\,cm^{-1}$ for $(HAc)_2$, but for the two isotope species it is expected to be smaller according to the anharmonic calculations at the B3LYP/6-311++G(2d,2p) level (see Tab. 4.7), about $2\,cm^{-1}$ $(DAc)_2$ and only $0.1\,cm^{-1}$ for $(D_4Ac)_2$. Therefore, it is relatively certain that the intense but slightly broad band of $(D_4Ac)_2$ at $1044\,cm^{-1}$ belongs to both modes which cannot be resolved. For the related two bands of $(DAc)_2$, there are three bands in the suitable wavenumber region in our jet spectrum, at 1433, 1440 and $1450\,cm^{-1}$, respectively. Depolarization analysis (see Fig. 4.9) shows that only the band at $1433\,cm^{-1}$ has A_g symmetry, which is therefore assigned to ν_5. The remaining two bands with B_g symmetry can be either from ν_{16} or from the combination band of $\nu_{10} + \nu_{19}$ predicted at $860 + 595 = 1455\,cm^{-1}$. A conceivable contribution from the ν_5 band of $(HAc)_2$ on the latter band can be neglected due to the different symmetry type, despite totally the same band positions ($1450\,cm^{-1}$). We prefer to assign the band at $1440\,cm^{-1}$ as ν_{16} and the other one as the combination band, which fits better to the calculations. The related combination band of $(HAc)_2$ is observed at $1497\,cm^{-1}$. Because the band is extremely weak, its symmetry cannot be safely determined with the depolarization experiment in Fig. 4.8. Weak anharmonicity constants are found: $-4\,cm^{-1}$ for $(HAc)_2$ ($1497 - 897 - 604 = -4\,cm^{-1}$) and $-5\,cm^{-1}$ for $(DAc)_2$ ($1450 - 860 - 595 = -5\,cm^{-1}$). Both bands are marked with Label D in the Fig. 4.4 and Fig. 4.5.

The intramolecular O–H out-of-plane bending mode (ν_{18}) of neither of the deuterated compounds is observed due to the extremely small Raman activity. This is not surprising because the related band of FAD is also very weak even with more than 3% acid in helium [4].

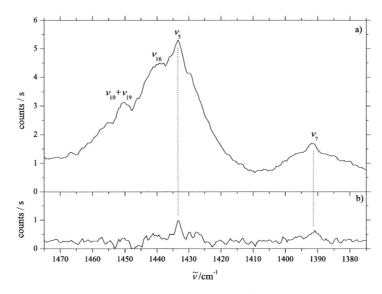

Figure 4.9: Spectrum of (DAc)$_2$ under **Condition I** (6×300 s) between 1360 and 1480 cm^{-1} with the excitation laser perpendicular to the scattering plane (top trace) and residual after subtracting 7/6 of the spectrum obtained with the excitation laser parallel to the scattering plane.

4.2.3 Non-fundamental modes

Beside the monomer bands there is a group of bands observed in Fig. 4.4 and Fig. 4.5 which do not correspond to the fundamental modes of AAD. They are marked with labels A-G and the same label in the different spectra means the same assignment.

Bands A and B are close to the intense C–C stretching band ν_{10} for all three species, more or less regularly and will be discussed at first. Fig. 4.10 shows a comparison of the three isotopomers in the ν_{10} region. In the low frequency wing of the main band, band A with a wavenumber shift of around 25 cm^{-1} is observed for all three species, whereas two bands marked with B are in the spectra of (DAc)$_2$ and (D$_4$Ac)$_2$ shifted about 15 cm^{-1} to the higher wavenumber relative

167

to the ν_{10} band. The case of band B is the same as in the measurement of the deuterated FAD [4]. The OCO in-plane bending modes of the unsymmetrically isotope-substituted dimers (DCOOH-DCOOD, HCOOH-HCOOD) were observed, both shifted by about $15\,\mathrm{cm}^{-1}$ to higher wavenumber relative to their symmetric dimer counterparts [4]. These mixed dimers can arise from partial isotope exchange at the container walls or from an isotopic impurity of the compound itself. DAc used in this work has a 98% isotopic purity (see Tab. 2.2). It is supported by the presence of the ν_{10} band of $(\mathrm{HAc})_2$ in the spectrum of $(\mathrm{DAc})_2$ at $897\,\mathrm{cm}^{-1}$ (see Fig. 4.10). Band B in the spectrum of $(\mathrm{D_4Ac})_2$ is not so clear as that in the spectrum of $(\mathrm{DAc})_2$ because the $\mathrm{D_4Ac}$ used in this work is 99.5% isotopically pure (see Tab. 2.2). Both blue-shifts are underestimated (see Tab. 4.8), but the assignment is still straightforward. Furthermore, the two bands which shift about

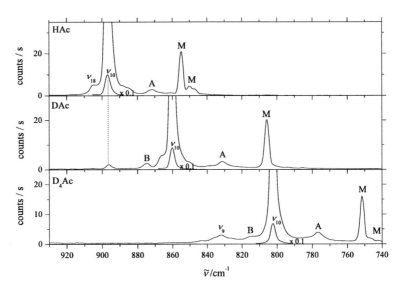

Figure 4.10: Raman spectra of the three isotopomers of AAD under **Condition I** in the C–C stretching vibration region in detail. The spectra are the same as those in Figs. 4.4 and 4.5

$25\,\mathrm{cm}^{-1}$ to the higher wavenumber of ν_4 on the spectra of $(\mathrm{DAc})_2$ and $(\mathrm{D_4Ac})_2$ (see Fig. 4.5, marked with Label G) should come from the same mechanism. Evidence of isotope exchange is not seen in the spectra of $(\mathrm{HAc})_2$, which were measured before introducing deuterated compounds into the apparatus.

Bands A come from the C–C stretching vibration of hetero AAD dimers too, involving an acid molecule and its ^{13}C isotopomer with a ^{13}C atom on the methyl group. Although the abundance of ^{13}C compounds is only about 1%, the band is already intense enough to be detected because of the large change in polarizability with C–C stretching motion. Wavenumber shifts of A relative to the ν_{10} band of its related homo dimer without ^{13}C atom (normal dimer) are nearly the same for the three isotopomers, because all the three normal dimers have a similar molecule mass and a similar wavenumber of C–C stretching vibration (see Tab. 4.8). The assignment is weakly supported by the calculations because the wavenumber shift

Dimer	Calculation	Experiment	Label
$(\mathrm{CH_3COOH})_2$	903.5(0/20.6)	897	
$(\mathrm{CH_3COOH})_2$	905.3(5.3/0)		
$\mathrm{CH_3COOH}\text{-}^{13}\mathrm{CH_3COOH}$	894.8(2.9/12.2)/904.7(3.5/7.8)	871 (-26)	A
$\mathrm{CH_3COOH}\text{-}\mathrm{CH_3}^{13}\mathrm{COOH}$	900.6(1.9/15.0)/904.9(4.3/5.5)		
$(\mathrm{CH_3COOD})_2$	865.2(0/16.4)	860	
$(\mathrm{CH_3COOD})_2$	866.9(8.1/0)		
$\mathrm{CH_3COOH}\text{-}\mathrm{CH_3COOD}$	866.3(3.8/9.0)/904.7(3.1/9.3)	875 (15)	B
$\mathrm{CH_3COOD}\text{-}^{13}\mathrm{CH_3COOD}$	855.3(4.3/9.4)/866.5(5.1/6.7)	832 (-28)	A
$\mathrm{CH_3COOD}\text{-}\mathrm{CH_3}^{13}\mathrm{COOD}$	863.8(2.1/12.9)/866.6(6.5/3.6)		
$(\mathrm{CD_3COOD})_2$	807.3(0/12.2)	802	
$(\mathrm{CD_3COOD})_2$	815.7(14.5/0)		
$\mathrm{CD_3COOD}\text{-}\mathrm{CD_3COOH}$	811.3(5.8/7.7)/851.5(5.1/5.8)	816 (14)	B
$\mathrm{CD_3COOD}\text{-}^{13}\mathrm{CD_3COOD}$	805.4(0.7/11.5)/814.8(14.9/0.5)	777 (-25)	A
$\mathrm{CD_3COOD}\text{-}\mathrm{CD_3}^{13}\mathrm{COOD}$	805.4(0.7/11.6)/814.6(14.3/0.6)		

Table 4.8: Comparison of the harmonic calculations at the B3LYP/6-31+G* level and experiment results (in cm^{-1}) of the C–C stretching vibrations of different isotopically substituted AAD dimers. No symmetry was used for the calculation of the hetero-dimers. The calculated IR intensities (in km/mol)/Raman activities (in Å^4/u) and the band shifts (in cm^{-1}) of isotopically mixed dimers relative to the corresponding symmetric dimer modes are listed in parentheses.

are much underestimated (see Tab. 4.8) but strongly supported by the fact that the band intensity ratios (ratios of the integrated areas of the bands) of A and the ν_{10} main band are nearly the same for all the three isotopomers (3%-4%). This is somewhat higher than the expected 2.2% abundance, in line with the predicted Raman activity which is more than 50% of that of the symmetric dimer.

As in the case of FAD, the lowest intramolecular mode of AAD is the strong OCO in-plane bending band ν_{12}, whereas all the intermolecular fundamentals are lower than half of its wavenumber. However, several strong overtone bands of the intermolecular modes were observed in the Raman spectra of FAD [4], but no similar bands were detected for AAD, even under **Condition II**. The reason of the much smaller intensity could be due to the higher mass of AAD.

More overtone/combination bands can be observed under **Condition II**, but also with the growing risk that these bands may come from larger cluster aggregates or IR active modes. Traces of mixed dimers make the assignment of the combination/overtone bands in the spectra of $(DAc)_2$ and $(D_4Ac)_2$ even more difficult. Here we only try to roughly assign the several groups of bands in the spectrum of $(HAc)_2$. The bands between ν_{17} and ν_8 of $(HAc)_2$ (marked with Label C) in Fig. 4.4 may correspond to the overtone/combination bands of ν_{11}, ν_{19}, ν_{26} and ν_{40}. Fermi resonance as well as further interactions are likely. Therefore, detailed assignments cannot be made, particularly also due to the lack of accurate IR fundamental band positions. This also applies to the assignment of the bands between 1475 and 1650 cm^{-1} (marked with Label D in Fig. 4.4) as well as the bands in the region of 1725-1800 cm^{-1} (Label: E). The former was considered as the combination bands of $\nu_{10} + (\nu_{11}/\nu_{19})$ or $\nu_{39} + (\nu_{40}/\nu_{26})$ and the latter as the overtone bands of ν_{10} and ν_{39}.

What should be mentioned here is the abnormal band shape of ν_4 and ν_8 on the spectrum of $(DAc)_2$ under **Condition I**. More than one shoulder was observed. Two intense broad monomer bands at $\sim$1660 cm^{-1} and $\sim$1330 cm^{-1} are observed in the Raman GP Spectrum (see Fig. 4.5, trace a)), which are very close to ν_4 and ν_8 of the dimer. Other sources of these sub bands could be from resonance with unknown combination bands or water-acid clusters.

4.3 Region above 1800 cm^{-1}: ν(O–H/D) and ν(C–H/D) vibrations

Raman jet spectra of the three AAD isotopomers between 1800 and 3200 cm^{-1} under dimer-dominant expansion conditions (**Condition I**) are shown in Fig. 4.11. The spectra of (HAc)$_2$ and (DAc)$_2$ are combinations of three measurements in different wavenumber regions under equivalent conditions. (D$_4$Ac)$_2$ was measured only up to $\sim$2800 cm^{-1} and the spectrum in Fig. 4.11 is a combination of two measurements. The C–H stretching vibration region is ten times reduced in the spectra of (HAc)$_2$ and (DAc)$_2$ and several bands of (DAc)$_2$ between 2350 and 2900 cm^{-1} are five times enlarged to fit the other bands and provide a better view.

Similar to FAD, the O–H/D stretching mode of AAD shows an irregular and complex band structure. For all the three substances, a group of bands was observed in a broad wavenumber region spread over about 400 cm^{-1}: 2500 to 2900 cm^{-1} for (HAc)$_2$ and 2000 to 2400 cm^{-1} for the two deuterated isotopomers. Unlike in the IR jet spectra (see Fig. 4.13), the C–H stretching bands are so pro-

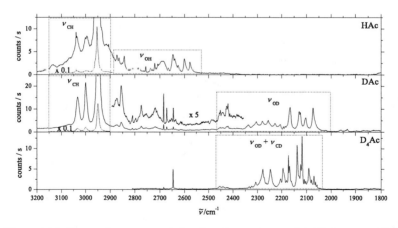

Figure 4.11: Raman jet spectra of the three isotopomers of AAD between 1800 and 3200 cm^{-1} under **Condition I**, 8×300 s.

nounced that they can be easily distinguished from the O–H stretching manifold. The O–D stretching mode has a relatively regular band structure and is significantly simpler in $(DAc)_2$ than in $(D_4Ac)_2$. The latter is more complicated due to the overlap with the three C–D stretching modes.

$(HAc)_2$ has a total of three Raman active C–H stretching vibrational modes: ν_2 and ν_3 with A_g as well as ν_{15} with B_g symmetry. Three related bands are observed in its jet spectrum, on the left side of Fig. 4.11, from about 2900 to $3200\,\mathrm{cm}^{-1}$. Quantum chemical calculations predict a wavenumber sequence of $\nu_2 > \nu_{15} > \nu_3$ (see Tab. 4.10 later on), whereas the depolarization experiment (see Fig. 4.12) shows that the two bands with higher wavenumbers have a high depolarization ratio. This is not surprising because ν_2 has a depolarization ratio $\rho_\perp = 0.73$ according to the harmonic calculation at the B3LYP/6-311++G(2d,2p) level despite its A_g symmetry. Therefore, the band at $3031\,\mathrm{cm}^{-1}$ is assigned as ν_2, the band at $2999\,\mathrm{cm}^{-1}$ in the middle as the asymmetric C–H stretching vibrational mode ν_{15} and the most intense band at $2948\,\mathrm{cm}^{-1}$ with the clear A_g symmetry character as ν_3. All the three C–H stretching vibration modes are also observed in the spectrum of $(DAc)_2$ with nearly completely the same band positions as those of $(HAc)_2$, and therefore with the same assignments. Deuteration of the hydroxyl group has almost no influence on the C–H stretching vibrations, which underlines the lack of coupling to the O–H modes. C–D stretching modes of $(D_4Ac)_2$ cannot be distinguished. The band positions and assignments are summarized in Tab. 4.9 and fit well with those of the former Raman GP study [128].

The main band of the O–H stretching vibrational region in the spectra of $(HAc)_2$ spreads from about 2500 to $2900\,\mathrm{cm}^{-1}$. There are some bands of $(DAc)_2$ in this wavenumber region too, but its main bands lie in the range between 2000 and $2400\,\mathrm{cm}^{-1}$, which is also the main O–D stretching vibration regions of $(D_4Ac)_2$. From the comparison of band positions we can take the bands of $(DAc)_2$ as the intermediate stage of the other two species: its bands between 2600 and $2900\,\mathrm{cm}^{-1}$ should have more methyl group vibration character because these bands disappear in the spectra of $(D_4Ac)_2$ with the deuteration of the methyl group; its bands between 2000 and $2400\,\mathrm{cm}^{-1}$ should have more hydroxyl group vibration character

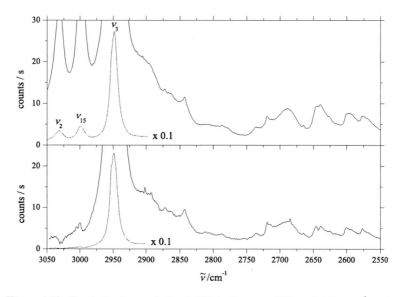

Figure 4.12: Depolarization analysis of (HAc)$_2$ between 2550 and 3050 cm^{-1} under **Condition II** (12 × 200 s), with the excitation laser perpendicular to the scattering plane (top trace) and residual after subtracting 7/6 of the spectrum obtained with the excitation laser parallel to the scattering plane. The C–H stretching region is ten times reduced to provide a better view.

Mode	Description	(HAc)$_2$ Jet[a]	(HAc)$_2$ GP[b]	(DAc)$_2$ Jet[a]	(D$_4$Ac)$_2$ Jet[a]
A$_g$					
ν_2	asym. ν(C–H/D)	3031	≈ 3035	3032	. . .
ν_3	sym. ν(C–H/D)	2948	2954.1	2949	. . .
B$_g$					
ν_{15}	asym. ν(C–H/D)	2999	. . .	3000	. . .

[a] Raman jet spectra, this work.
[b] Bertie and Michaelian, Raman GP spectrum, Ref. 128

Table 4.9: Band maxima (in cm^{-1}) and assignment of the three Raman active C–H/D stretching fundamentals of (HAc)$_2$ and its OD/D$_4$ isotopomers, compared with the former GP study.

Mode	$(HAc)_2$	$(DAc)_2$	$(D_4Ac)_2$
A_g			
ν_2	3167.5(0.73)/3015.1	3167.2(0.69)/3015.3	2348.7(0.75)/2266.8
ν_3	3060.5(0.02)/2932.7	3060.8(0.01)/2943.0	2199.8(0.01)/2100.3
B_g			
ν_{15}	3116.1(0.75)/2963.4	3116.1(0.75)/2964.2	2305.2(0.75)/2220.6

Table 4.10: Comparison of harmonic/anharmonic calculations of the three Raman active C–H stretching fundamentals of $(HAc)_2$ and its OD/D_4 isotopomers at the B3LYP/6-311++G(2d,2p) level (in cm^{-1}). The depolarization ratios $\rho_\perp$ from the harmonic calculations are listed in parentheses to assist the band assignments.

because these bands appear only in the spectra of $(D_4Ac)_2$ but disappear in the spectra of $(HAc)_2$.

Although all the Raman active vibrations in the lower wavenumber region have already been measured and assigned in this work, it is still a great challenge to assign the bands in the O–H/D stretching vibrational regions. The bands are more regular than those of FAD, but there are also more options of assignments due to more numerous finger print bands of AAD compared to FAD. More information can be obtained from comparisons between Raman and IR spectra, because a good correspondence can be found between the band positions of the Raman/IR jet spectra of $(HAc)_2$ and $(DAc)_2$ in the O–H/D stretching vibrational region. Parallel comparisons of the Raman/IR spectra of $(HAc)_2$ and its OD/D_4 isotopomers are shown in Figs. 4.13, 4.14 and 4.15. All the IR spectra are taken from Ref. 38 and measured with a concentration of 0.23% acid in helium.

It is interesting to see that the IR active C–H stretching modes of $(DAc)_2$ are not observed in the IR jet spectra (see Fig. 4.14), in contrast to the case of the very intense Raman active C–H bands. Actually, the IR active C–H stretching bands of $(HAc)_2$ are also relatively weak, even with contributions from the interaction with the O–H stretching mode (see Fig. 4.13). These different band intensities can be straightforwardly explained with the calculated IR activities of the C/O–H/D stretching modes (see Tab. 4.11). Firstly, the IR activities of all C–H/D

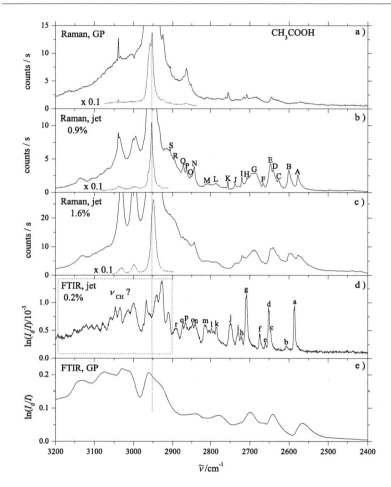

Figure 4.13: Comparison of the Raman/IR spectra of (HAc)$_2$ between 2400 and 3200 cm^{-1}. All the Raman spectra are combinations of two measurements in different wavenumber regions under equivalent conditions. The C–H stretching region is ten times reduced to provide a better view. a) Raman GP spectrum, $T_s = 20\,°C$ (1.6% HAc in He), $p_b = 30\,mbar$, $4 \times 60\,s$; b) Raman jet spectrum, **Condition I**, $8 \times 300\,s$; c) Raman jet spectrum, **Condition II**, $8 \times 300\,s$ ($8 \times 200\,s$ for the spectrum between 2800 and 3200 cm^{-1}); d) FTIR jet spectrum [38]; e) FTIR GP spectrum [38].

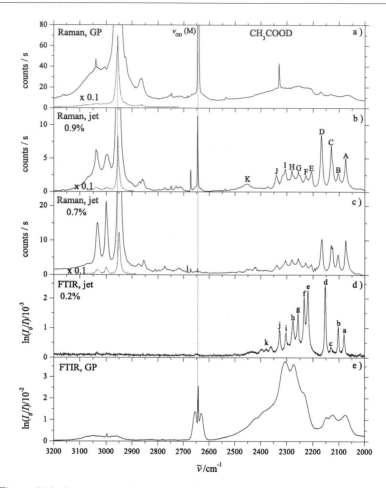

Figure 4.14: Comparison of the Raman/IR spectra of $(DAc)_2$ between 2000 and 3200 cm^{-1}. All the Raman spectra are combinations of three measurements in different wavenumber regions under equivalent conditions. The C–H stretching region is ten times reduced to provide a better view. a) Raman GP spectrum, $T_s = 8°C$ (0.6% DAc in He), $p_b = 30$ mbar, 4×30 s; b) Raman jet spectrum, **Condition I**, 8×300 s; c) Raman jet spectrum, $T_s = 8°C$ (0.6% DAc in He), $p_s = 700$ mbar, $d = 1$ mm, 8×300 s; d) FTIR jet spectrum [38] ; e) FTIR GP spectrum [38].

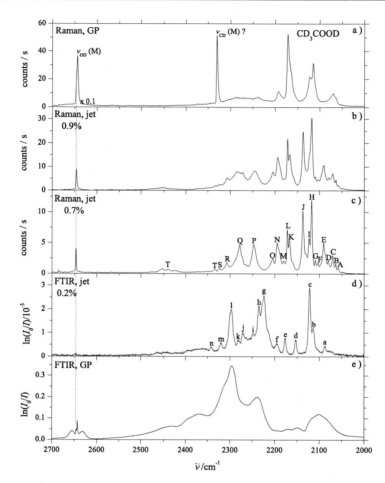

Figure 4.15: Comparison of the Raman/IR spectra of (D$_4$Ac)$_2$ between 2400 and 3200 cm^{-1}. All the Raman spectra are combinations of two measurements in different wavenumber regions under equivalent conditions. a) Raman GP spectrum, $T_s = 8°$C (0.6% D$_4$Ac in He), $p_b = 30$ mbar, 4 × 30 s; b) Raman jet spectrum, **Condition I**, 8 × 300 s; c) Raman jet spectrum, $T_s = 8°$C (0.6% D$_4$Ac in He), $p_s = 700$ mbar, $d = 1$ mm, 8 × 300 s; d) FTIR jet spectrum [38]; e) FTIR GP spectrum [38].

fundamentals are much smaller than that of the O–H/D stretching mode for all the three isotopomers. The C–H stretching mode of $(HAc)_2$ with the highest IR activity is about six times weaker after the deuteration of the hydroxyl group, indicating some O–H mixing. The IR activities of the C–D stretching modes are even weaker than those of $(DAc)_2$ after the deuteration of the methyl group (see Tab. 4.11) and so the contributions on the IR active O-D stretching band intensities should be weakened furthermore. That may explain why the IR active bands are less visible than the Raman ones in the C/O–D stretching region of $(D_4Ac)_2$ and the correspondence between the band positions of the Raman/IR jet spectra is less pronounced.

The discussions on the complicated O–H/D bands are started with $(HAc)_2$. The Raman active overtone/combination bands are marked with labels A-S and the IR active ones with labels a-s in Fig. 4.13. All the Raman active bands between 2550 and 2900 cm^{-1} have A_g symmetry, determined by the depolarization analysis shown in Fig. 4.12, whereas the polarized IR spectra of $(HAc)_2$ indicate that the IR active bands between 2500 and 3200 cm^{-1} are mainly of B_u symmetry [133]. Beside that, the band positions of the Raman/IR bands marked with the same label are nearly the same. All the evidence shows that they are Raman/IR active "corresponding bands" and most of the bands should come from binary combinations. The change of only one Davydov partner band by the combination leads to the small wavenumber differences between the corresponding Raman/IR

Mode	Description	$(HAc)_2$	$(DAc)_2$	$(D_4Ac)_2$
B_u				
ν_{30}	ν(O–H/D)	3190(3263.3)/2857	2327(1745.6)/2115	2327(1744.9)/2099
ν_{31}	sym. ν(C–H/D)	3167(57.9)/3014	3167(9.6)/3015	2349(4.1)/2267
ν_{32}	sym. ν(C–H/D)	3061(2.2)/2938	3061(2.4)/2943	2200(0.4)/2104
A_u				
ν_{22}	asym. ν(C–H/D)	3116(6.0)/2963	3116(6.0)/2964	2305(2.9)/2221

Table 4.11: Comparison of harmonic/anharmonic calculations of the IR active C/O–H/D stretching fundamentals of $(HAc)_2$ and its OD/D_4 isotopomers at the B3LYP/6-311++G(2d,2p) level (in cm^{-1}). The IR activities from the harmonic calculations are listed in parentheses.

combination bands. The band positions and the assignments from several former studies are summarized in Tab. 4.12. Assignments of the IR active bands are from J. Dreyer [17,129], who studied the anharmonic coupling in this region with density-functional theory and made simulations of the IR spectra of $(HAc)_2$, which fit the experimental results very well.

Band A at 2577 cm^{-1} in the Raman jet spectra of $(HAc)_2$ is shown as a single, broad band in the gas phase at $\sim$2571 cm^{-1} (see Fig. 4.13), which was observed at $\sim$2565 cm^{-1} by Bertie *et al.* [128] in the Raman GP spectra and assigned as the first overtone of the A_g-symmetrical C–O-stretching vibration ν_8. This fundamental is observed at 1294 cm^{-1} in the jet spectrum (this work) and at 1285 cm^{-1} in the GP measurement [128]. Its B_u-symmetric Davydov partner ν_{37} is determined at 1307 cm^{-1} in FTIR-jet and at 1294 cm^{-1} in the GP [38]. The Davydov splitting between them in the GP is about 9 cm^{-1}, which does not change substantially (13 cm^{-1}) under expansion conditions.

If band A was assigned as the overtone band of ν_8 (2×1294 cm$^{-1} = 2588$ cm^{-1}), it is logical to assign band B at 2600 cm^{-1} in the Raman jet spectrum as the overtone of ν_{37} (2×1307 cm$^{-1} = 2614$ cm^{-1}). The fundamental band positions observed in the Raman jet measurement under **Condition I** and in the FTIR jet spectra [38] (see Tab. 4.5) are used for both predictions and will be used for the further ones. Then theoretically there should be a band in the IR spectrum, which is expected to lie exactly between bands A and B. The distance between A and B should be roughly twice the Davydov splitting of the fundamentals from the jet measurements (2×13 cm$^{-1} = 26$ cm^{-1}). Both expectations are true. The wavenumber difference between A and B is 23 cm^{-1} and the IR spectrum shows a band at 2585 cm^{-1} [38], which can be assigned as the combination band of ν_8 and ν_{37}. Emmeluth *et al.* [138] have made a corresponding assignment of the IR bands based on the band shifts of fundamental bands with Argon coating experiment. The shift of the band at 2585 cm^{-1} is roughly twice the displacement of the fundamental band ν_{37}. Anharmonic quantum mechanical calculations at the B3LYP/6-311+G(d,p) level are also in accordance with this prediction [17]. An in-

	Raman active				IR active		
Label	Jet[a]	GP[b]	Assignment[b]	Label	Jet[c]	Simulation[d]	Assignment[d]
A	2577	~2565	$2\nu_{C-O}(A_g)$	a	2585	2582	$\nu_{C-O}(A_g) + \nu_{C-O}(B_u)$
B	2600	...		b	2606	...	...
C	2626	...		c	2646	2647	$\nu_{C-O}(A_g) + \delta_s(CH_3)(B_u)$
D	2640	2635	$\nu_{C-O}(A_g) + \delta_s(CH_3)(A_g)$	d	2652	2647	$\nu_{C-O}(A_g) + \delta_s(CH_3)(B_u)$
E	2647	...		e	2660	...	...
F	2666	...		f	2675	2677	$\delta_s(CH_3)(A_g) + \nu_{C-O}(B_u)$
G	~2691	2681	$\nu_{C=O}(A_g) + \rho_s(CH_3)(A_g)$	g	2708	2704	$\nu_{C=O}(A_g) + \nu_{C-O}(B_u)$
H	2708	...		h	2722	2728	$\nu_{C-O}(A_g) + \nu_{C-O}(B_u) + \delta_{O-H}(B_u)$
I	2720	...		i	2730	2730	$\delta_{O-H}(A_g) + \nu_{C-O}(B_u)$
J	2737	...		j	2749	2751	$\nu_{C-O}(A_g) + \nu_{C-O}(B_u) + \nu_{O-H\cdots O}(A_g)$
K	2756	...		k	2785	2784	$\delta_s(CH_3)(A_g) + \delta_{O-H}(B_u)$
L	~2787	2771	$\delta_s{}'(CH_3)(A_g) + \delta_s(CH_3)(A_g)$	l	2798	2791	$\delta_{O-H}(A_g) + \delta_s(CH_3)(B_u)$
M	~2809	...		m	~2815	2816	$\delta_s(CH_3)(A_g) + \nu_{C-O}(B_u) + \delta_{O-H\cdots O}(A_g)$
N	2843	...		n	2841	2822	$\nu_{C-O}(A_g) + \delta_s(CH_3)(B_u) + \nu_{O-H\cdots O}(A_g)$
O	2852	...		o	2848	2841	$\delta_{O-H}(A_g) + \delta_{O-H}(B_u)$
P	2864	...		p	2867	2849	$\nu_{C-O}(A_g) + \delta_{O-H}(B_u) + \delta_{O-H}(B_u)$
Q	2873	2865	$\delta_s{}'(CH_3)(A_g) + \delta_{O-H}(A_g)$	q	2874	2872	$\nu_{C-O}(A_g) + \delta_{O-H}(B_u) + \nu_{O-H\cdots O}(A_g)$
R	2895	...		r	~2891	2897	$\delta_{O-H}(A_g) + \nu_{C-O}(B_u) + \nu_{O-H\cdots O}(A_g)$
S	2906	...					

[a] Raman jet spectrum, this work.
[b] Bertie and Michaelian, Raman GP spectrum, Ref. 128.
[c] Häber, FTIR jet spectrum, shown in Ref. 38.
[d] Dreyer, simulated IR spectrum, Ref. 17.

Table 4.12: Band positions (in cm^{-1}) and assignments of the overtone/combination bands of (HAc)$_2$ in Fig. 4.13, compared with some former studies. See text for more details.

tense combination band dominated by $\nu_8 + \nu_{37}$ lies at 2582 cm^{-1} in the theoretical simulated spectrum (see Tab. 4.12).

Band D at 2640 cm^{-1} in our jet spectrum was observed at 2635 cm^{-1} in the GP spectrum [128] and assigned as the combination band of $\nu_7 + \nu_8$ (see Tab. 4.12). The combination band of the B_u-symmetric Davydov partners of the two fundamentals is $\nu_{36} + \nu_{37}$ (1363 cm^{-1} + 1306/1308 cm^{-1} = 2669/2671 cm^{-1}), which is 4/6 cm^{-1} higher than $\nu_7 + \nu_8$ (1371 cm^{-1} + 1294 cm^{-1} = 2665 cm^{-1}) and should be the band E at 2647 cm^{-1}. The two corresponding IR active combination bands are assumed at 2657 cm^{-1} ($\nu_8 + \nu_{36}$, 1294 cm^{-1} + 1363 cm^{-1} = 2657 cm^{-1}) and 2677/2679 cm^{-1} ($\nu_7 + \nu_{37}$, 1371 cm^{-1} + 1306/1308 cm^{-1} = 2677/2679 cm^{-1}), which fit very well to the band d at 2652 cm^{-1} and the band f at 2675 cm^{-1} in the FTIR jet spectrum in Fig. 4.13. Two simulated bands at 2647 cm^{-1} and 2677 cm^{-1} with the same assignments were shown from the theoretical work [17] (see Tab. 4.12).

For all the combination bands discussed above, both Raman and IR active, small "anharmonicity constants" ($x_{i,j}$) within 25 cm^{-1} are found. No further discussions are made for the combination bands with higher wavenumbers due to the presence of ternary combinations according to the simulation.

The band positions of the Raman/IR combination bands of the two deuterated isotopomers are listed in Tabs. 4.13 and 4.14 but without assignments. The depolarization analyses show that most Raman active combination bands have A_g symmetry except the two intense bands P and Q of (D$_4$Ac)$_2$ (see Figs. 4.16 and 4.17).

Similar to the O-deuterated formic acid (see Fig. 3.37), the jet spectra of (DAc)$_2$ between 2200 and 2400 cm^{-1} show a regular pattern with six band maxima (Raman: E to J; IR, e to j, see Fig. 4.14 and Tab. 4.13). They shift to higher wavenumbers in the GP spectrum (i.e. weakened hydrogen bonds), indicating a strong O–H stretching character [33]. In contrast, the bands between 2000 and 2200 cm^{-1} (Raman: A to D; IR, a to d, see Fig. 4.14 and Tab. 4.13) are red-shifted in the gas phase. This is an indication for C–O stretching vibration character as C–O bonds acquire a partial double bond character with increasing strength of

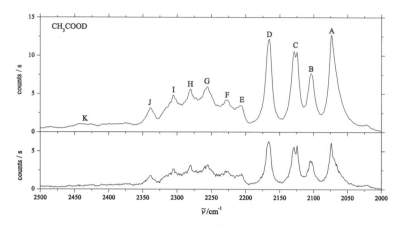

Figure 4.16: Depolarization analysis of (DAc)$_2$ between 2000 and 2500 cm^{-1}, with the excitation laser perpendicular to the scattering plane (top trace) and residual after subtracting 7/6 of the spectrum obtained with the excitation laser parallel to the scattering plane. $T_s = 8°C$ (0.6% DAc in He), $p_s = 700$ mbar, $d = 1$ mm, 8×300 s. The labels are the same as those used in Fig. 4.14.

Raman active		IR active	
Label	Jet^a	Label	Jet^b
A	2074	a	2080
B	2103	b	2103
C	2130	c	2131
D	2168	d	2152
E	2206	e	2219
F	2227	f	2233
G	2257	g	2257
H	2281	h	2277
I	2306	i	2303
J	2339	j	2328
K	~2453	k	2362/2380/2398

^{a} Raman jet spectrum, this work.
^{b} Häber, FTIR jet spectrum, shown in Ref. 38.

Table 4.13: Band positions (in cm^{-1}) of the overtone/combination bands of (DAc)$_2$ in Fig. 4.14.

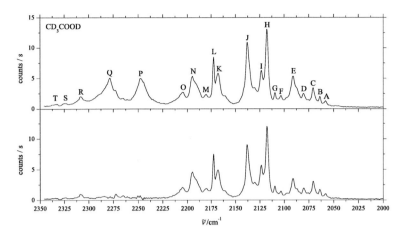

Figure 4.17: Depolarization analysis of $(D_4Ac)_2$ between 2000 and $2350\,cm^{-1}$, with the excitation laser perpendicular to the scattering plane (top trace) and residual after subtracting 7/6 of the spectrum obtained with the excitation laser parallel to the scattering plane. $T_s = 8°C$ (0.6% D_4Ac in He), $p_s = 700\,mbar$, $d = 1\,mm$, $8 \times 300\,s$. The labels are the same as those used in Fig. 4.15.

the hydrogen bonds under the jet conditions and therefore have a larger wavenumber. Both O-deuterated formic acid and acetic acid should be excellent model systems for detailed theoretical studies. The bands of the experimental spectra are sufficiently separated from each other, with precisely determined band positions.

There are more bands in the O–D stretching fundamental region in the jet spectra of $(D_4Ac)_2$ between 2000 and $2400\,cm^{-1}$ than in $(DAc)_2$. Several bands of these two compounds have similar band positions and intensities and can be regarded as bands from the same combination partners.

4.4 Conclusions

In this chapter, the Raman spectra of $(HAc)_2$ and its OD/D_4 isotopomers are discussed in the entire fundamental wavenumber region, mainly focused on the

Raman active		IR active	
Label	Jet[a]	Label	Jet[b]
A	2058	a	2088
B	2064	b	2114
C	2071	c	2122
D	2081	d	2154
E	2091	e	2177
F	2103	f	2196
G	2109	g	2224
H	2118	h	2235
I	2123	i	2249
J	2138	j	2271
K	2168	k	2282
L	2172	l	2297
M	2180	m	2320
N	2196	n	2342
O	2205		
P	2247		
Q	2279		
R	2308		
S	2325		
T	2332		

[a] Raman jet spectrum, this work.
[b] Häber, FTIR jet spectrum, shown in Ref. 38.

Table 4.14: Band positions (in cm^{-1}) of the overtone/combination bands of $(D_4Ac)_2$ in Fig. 4.15.

fundamental modes. They are assigned and compared with former experimental results as well as with quantum chemical calculations. Several modes, e.g. the in-plane bending mode of isolated acetic acid dimer, are assigned correctly for the first time and major corrections to previous thermalized gas phase values were derived for the other Raman active hydrogen bond modes as well. The C=O stretching fundamental shows a simple single band in the Raman jet spectra of all three isotopomers, not like the case of FAD. Less combination/overtone bands of AAD were observed in the jet spectra, but their assignment is less certain than that of FAD due to more combination possibilities from the more numerous fundamental modes of AAD in the low wavenumber region. Only a few of them are assigned, with help of the IR spectra measured by Häber [38] in the O/C–H/D fundamental region. The C–C stretching bands from mixed dimers are analyzed.

The influence of experimental conditions on the band shapes and wavenumbers of AAD is discussed in detail in different wavenumber regions. It is shown that supra-dimer aggregates which form in concentrated acetic acid-seeded expansions differ from isolated dimers in their out-of-plane modes, thus indicating a stacking geometry of these supramolecular units. On a methodical level, it is demonstrated that Raman spectra of carboxylic acid dimers cooled in supersonic jets and also partially re-heated by thermalization shock waves can be recorded in the low frequency hydrogen bond mode region, arguably the most important spectral window for the statistical thermodynamics of dimer formation.

5 Pivalic and Propiolic acid

By tuning the alkyl substituent of a carboxylic acid, the differences on the vibrational properties and the electronic effects of such substitutions on the strength of the double hydrogen bond can be compared and analyzed. The formic acid and acetic acid have been discussed in Chaps. 3 and 4. In this chapter, we investigate the substitution of the C-attached hydrogen in formic acid specifically by an ethynyl (propiolic acid) and a *t*-butyl group (pivalic acid) using Raman jet spectroscopy. In the first part of the chapter, the discussion of the four acids (formic, acetic, propiolic and pivalic acid) will be focused on intermolecular modes in their hydrogen-bonded dimers. Analysis of the dimer stretching mode within a pseudo-diatomic model quantifies substitution effects on the hydrogen-bond force constant. They do not correlate with predicted overall dimer binding energies. Furthermore, the Raman active intramolecular fundamentals of these two acids will be discussed, compared with quantum chemical calculations and former work.

Both acids form centrosymmetric (C_{2h}) cyclic dimers. Due to this nature, both Raman (A_g and B_g modes) and IR spectra (A_u and B_u modes) are needed for a complete description. Technically, panoramic supersonic jet spectra in the relevant wavenumber range below $300\,\text{cm}^{-1}$ are still largely unfeasible in the infrared case [11, 16], whereas they have recently been demonstrated in the case of spontaneous Raman scattering [3, 4, 26]. Furthermore, the important dimer stretching mode is Raman active and so is one member each of both Davydov pairs related to in-plane and out-of-plane bending motion. Therefore, the present study focuses on the Raman active modes, but discusses briefly the implications for IR modes in thermally excited or embedded dimers.

5.1 Comparison of the intermolecular modes of the four acid dimers

5.1.1 Effects on the band positions

Fig. 5.1 shows spontaneous Raman scattering spectra recorded for expansions of propiolic acid and pivalic acid in the 275-75 cm^{-1} range, compared with those of formic acid and acetic acid. Conditions were chosen such that the expansion is dominated by dimers, with smaller contributions by monomers and larger clusters. The large changes in appearance are mostly due to the mass effect of the organic group attached to the -COOH unit. The dimer stretching mode $\nu(A_g)$ decreases systematically in frequency with increasing mass, as expected. It exhibits a relatively sharp Q-branch with the exception of propiolic acid dimer. The in-plane bending mode δ shows a similar mass dependence and may always be found somewhat below the stretching mode. The out-of-plane bending mode γ shows a broader Raman band and drops more sharply with increasing substitution, such that it is out of the measurement range for pivalic acid dimer. Finally, the substituted carboxylic acid dimers have intramonomer modes in the displayed frequency window, namely torsional (τ) and C-C deformation modes (Γ, Δ).

In order to obtain these assignments, low level quantum chemical approaches are usually sufficient, because the coarse trends are dominated by simple mass and geometry effects. Tab. 5.1 shows such inexpensive predictions. The success of the harmonic MP2/6-31+G* approach in matching experimentally observed anharmonic intermolecular modes has been emphasized before [4] and was shown to be due to rather systematic error compensation, which works well except for the dimer stretching mode with its peculiar anharmonic zero point energy weakening effect [3] (see Tab. 5.2). Tab. 5.1 shows that it also works exceptionally well for the other carboxylic acid dimers. The larger basis set MP2 results are definitely too soft for the γ mode. The B3LYP results are less basis-set dependent and systematically higher than experiment, in line with expected anharmonic effects [4]. A dispersion-corrected density functional (B97D [139]) yields rather similar values, as expected for widely separated substituents.

The additional effect of six methyl groups on the Raman spectrum in the case of pivalic acid dimer shall be discussed. The out-of-plane mode γ is shifted out of the accessible spectral range, whereas the in-plane stretching and bending modes ν and δ move closer together and shift further down. The stretching mode is expected to decrease by a factor of $\sqrt{10/17} = 0.77$ in the harmonic limit, the actual factor is 0.73. This reflects a significantly softer hydrogen bond in pivalic acid dimer than in acetic acid dimer.

A more complex situation arises for propiolic acid dimer, where one would expect the stretching mode at $160 \pm 5\,\mathrm{cm}^{-1}$, by analogy to the other dimers. There

		MP2		B3LYP		B97D
Mode	Exp.	6-31+G*	6-311+G*	6-31+G*	6-311++G(2d,2p)	TZVP
(HCOOH)$_2$						
γ(O–H...O) (B$_g$)	242	242.3	209.0	255.1	261.8	260.9
δ(O–H...O) (A$_g$)	161	162.1	159.1	169.5	173.7	177.4
ν(O–H...O) (A$_g$)	194	202.9	191.6	209.3	207.1	202.4
(CH$_3$COOH)$_2$						
γ(O–H...O) (B$_g$)	111	111.5	92.0	124.0	123.2	120.1
δ(O–H...O) (A$_g$)	152	152.2	158.8	159.3	162.0	166.2
ν(O–H...O) (A$_g$)	173	177.0	168.3	182.1	179.9	176.9
(HCCCOOH)$_2$						
γ(O–H...O) (B$_g$)	85	84.7	55.3	89.8	87.0	81.0
δ(O–H...O) (A$_g$)	121	122.2	122.3	127.6	129.7	128.8
ν(O–H...O) (A$_g$)	152	159.5	154.2	163.2	157.7	159.1
Δ(C–C–C) (M^a)	181	176.5	184.7	191.8	190.4	184.7
Δ(C–C–C) (A$_g$)	214	215.2	216.1	228.0	227.6	225.3
Γ(C–C–C) (M^a)	227	223.1	235.7	264.5	242.0	219.1
Γ(C–C–C) (B$_g$)	254	257.4	243.7	297.4	270.3	244.6
(C(CH$_3$)$_3$COOH)$_2$						
γ(O–H...O) (B$_g$)	< 80	78.9	...	82.5	82.5	81.6
δ(O–H...O) (A$_g$)	118	119.9	...	123.7	122.5	130.0
ν(O–H...O) (A$_g$)	126	129.5	...	131.2	129.9	127.2
τ(CH$_3$) (B$_g$)	206	201.8	...	209.6	209.7	209.0
τ'(CH$_3$) (B$_g$)	245	245.6	...	251.5	250.8	249.3

^{a} Monomer.

Table 5.1: Comparison of calculated harmonic low frequency fundamentals of the four carboxylic acids using different quantum chemical methods and basis sets with experimental values (in cm^{-1}) from Fig. 5.1.

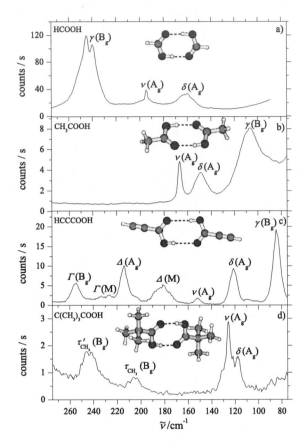

Figure 5.1: Raman jet spectra for expansions of He seeded with a) formic acid, b) acetic acid, c) propiolic acid and d) pivalic acid in the intermolecular fundamental range. Spectrum a) is the same as that in Fig. 3.12; b) is the same as that in Fig. 4.1; c) was taken under following conditions: $T_s = 16°C$ ($< 0.5\%$ HCCCOOH in He), $p_s = 1000$ mbar, $d = 0.4$ mm, 12×300 s, 8.0×0.05 mm^2 nozzle, 67 L reservoir, 250 m^3/h Roots pump and 100 m^3/h rotary vane pump; d) $T_s = 24°C$ ($\sim 0.07\%$ C(CH$_3$)$_3$COOH in He), $p_s = 800$ mbar, $d = 0.5$ mm, 12×300 s. Dimer stretching vibrations are marked ν, in-plane bending modes δ, out-of-plane bending modes γ, torsion modes of the methyl group τ, in-plane C–C–C deformation vibrations of the propiolic acid Δ and out-of-plane C–C–C deformation modes Γ. M indicates monomer signals.

Mode	Experiment	MP2/6-31+G*	Deviation
$(HCOOH)_2$			
$\gamma(O\text{–}H\ldots O)$	242	242	0
$\delta(O\text{–}H\ldots O)$	161	162	0.6%
$\nu(O\text{–}H\ldots O)$	194	203	4.6%
$(DCOOH)_2$			
$\gamma(O\text{–}H\ldots O)$	212	212	0
$\delta(O\text{–}H\ldots O)$	159	161	1.3%
$\nu(O\text{–}H\ldots O)$	193	200	3.6%
$(HCOOD)_2$			
$\gamma(O\text{–}D\ldots O)$	238	237	−0.4%
$\delta(O\text{–}D\ldots O)$	157	158	0.6%
$\nu(O\text{–}D\ldots O)$	194	202	4.1%
$(DCOOD)_2$			
$\gamma(O\text{–}D\ldots O)$	210	209	−0.5%
$\delta(O\text{–}D\ldots O)$	157	158	0.6%
$\nu(O\text{–}D\ldots O)$	192	199	3.6%
$(CH_3COOH)_2$			
$\gamma(O\text{–}H\ldots O)$	111	112	0.9%
$\delta(O\text{–}H\ldots O)$	152	152	0
$\nu(O\text{–}H\ldots O)$	173	177	2.3%
$(CH_3COOD)_2$			
$\gamma(O\text{–}D\ldots O)$	111	111	0
$\delta(O\text{–}D\ldots O)$	151	150	−0.7%
$\nu(O\text{–}D\ldots O)$	172	176	2.3%
$(CD_3COOD)_2$			
$\gamma(O\text{–}D\ldots O)$	106	104	−1.9%
$\delta(O\text{–}D\ldots O)$	140	147	1.3%
$\nu(O\text{–}D\ldots O)$	167	171	2.4%
$(HCCCOOH)_2$			
$\gamma(O\text{–}H\ldots O)$	85	85	0
$\delta(O\text{–}H\ldots O)$	121	122	0.8%
$\nu(O\text{–}H\ldots O)$	152	160	5.3%
$(C(CH_3)_3COOH)_2$			
$\gamma(O\text{–}H\ldots O)\ (B_g)$	< 80	79	...
$\delta(O\text{–}H\ldots O)\ (A_g)$	118	120	1.7%
$\nu(O\text{–}H\ldots O)\ (A_g)$	126	130	3.2%

Table 5.2: Comparison of calculated harmonic low frequency fundamentals of several carboxylic acids at the MP2/6-31+G* level with experimental values (in cm^{-1}). The relative deviation of the two corresponding values are also given as $(\tilde{\nu}_{\text{cal.}} - \tilde{\nu}_{\text{exp.}})/\tilde{\nu}_{\text{exp.}}$.

is an anomalously weak band at $152\,\mathrm{cm}^{-1}$ and a stronger band at $181\,\mathrm{cm}^{-1}$. The latter finds a straightforward explanation as a monomer C–C–C in-plane bending mode Δ and its stronger dimeric counterpart is indeed found above $200\,\mathrm{cm}^{-1}$. A similar pattern of weaker monomer and stronger blue-shifted dimer band is found for the out-of-plane C–C–C bending mode Γ (see also Tab. 5.1). The presence of monomer signals is due to the relatively dilute expansion. After having explained the intramolecular vibrations, the dimer stretching mode must correspond to the $152\,\mathrm{cm}^{-1}$ signal and indicates a softer hydrogen bond in propiolic acid dimer.

5.1.2 Stretching force constant and dissociation energy trends

The availability of experimental dimer stretching fundamental wavenumbers $\tilde{\nu}$ for all four carboxylic acid dimers invites a detailed analysis of substitution trends. Tab. 5.3 and Fig. 5.2 summarize the effective diatomic model force constants f extracted from

$$f = 4\pi^2 c^2 \tilde{\nu}^2 \mu$$

by using one half of the monomer masses as the appropriate reduced mass μ for both experiment and theoretical harmonic calculations. The species are arranged in the order of expected inductive effects, with electron-donating substituents (*t*-butyl, methyl) to the left and the presumably electron-withdrawing (ethynyl) substituent to the right of formic acid. One can see that the experimental force constant (green squares) drops from acetic acid over formic acid to propiolic acid, but also to pivalic acid, by about 10%. A straightforward interpretation is that acetic acid provides the best compromise among hydrogen bond acceptor strength of the C=O group and hydrogen bond donor strength of the O–H group for strong hydrogen bonding. A more strongly electron donating group R enhances the acceptor quality of the C=O group, but at the same time deteriorates the donor quality of the O–H group to a larger extent. The opposite is the case for a less strongly electron donating group. It would be quite difficult to make such a prediction by chemical "intuition", whereas the present experimental data leave

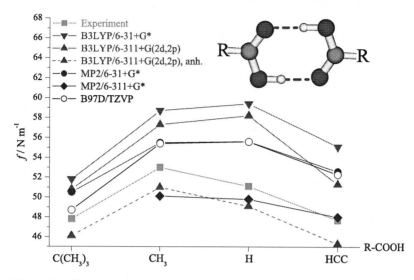

Figure 5.2: Comparison of experimental and calculated model force constant trends in the dimer stretching vibration ν for four carboxylic acids.

little room for another interpretation. Acetic acid dimer has the stiffest hydrogen bonds.

Is this finding confirmed by electronic structure calculations? The answer is no, as long as harmonic force constants are investigated. At B3LYP level, formic acid dimer is slightly stiffer than acetic acid dimer, whereas the further trends are qualitatively reproduced. At B97D and MP2 levels, formic and acetic acid dimer have essentially the same harmonic force constant. Again, the qualitative trends beyond these smallest dimers are reproduced. A major improvement results from perturbative inclusion of anharmonic contributions [115] in the calculation of the stretching fundamental (triangles connected by the dashed line). Apart from a systematic offset by about 4%, it reproduces the experimental trend very well. This is reminiscent of the findings for formic acid dimer [4, 69], where anharmonic B3LYP/6-311++G(2d,2p) calculations provided the most systematic agreement.

| Dimer | Experiment | | MP2 | | | | B3LYP | | | | B97D | |
| | ν | f | 6-31+G* | | 6-311+G* | | 6-31+G* | | 6-311++G(2d,2p) | | TZVP | |
			ν	f	ν	f	ν	f	ν	f	ν	f
(HCOOH)₂	194	51.1	202.9	55.9	191.6	49.8	209.3	59.4	207.1 (190.2)	58.2 (49.1)	202.4	55.6
(DCOOH)₂	193	51.7	200.4	55.7	189.3	49.7	206.8	59.3	204.7 (188.4)	58.1 (49.2)	...	...
(HCOOD)₂	194	52.2	201.7	56.4	190.4	50.3	208.1	60.0	206.2 (191.1)	58.9 (50.6)	...	...
(DCOOD)₂	192	52.2	199.3	56.2	188.1	50.1	205.6	59.9	203.8 (188.8)	58.9 (50.5)	...	...
(CH₃COOH)₂	173	53.0	177.0	55.5	168.3	50.1	182.1	58.7	179.9 (169.7)	57.3 (51.0)	176.9	55.4
(CH₃COOD)₂	172	53.3	175.9	55.7	167.1	50.3	181.0	59.0	178.7 (167.3)	57.5 (50.4)	...	...
(CD₃COOD)₂	167	52.7	170.6	55.0	162.2	49.7	175.4	58.2	173.7 (161.0)	57.0 (49.0)	...	...
(HCCCOOH)₂	152	47.7	159.5	52.6	152.4	48.0	163.3	55.1	157.7 (148.2)	51.3 (45.3)	159.1	52.3
(C(CH₃)₃COOH)₂	126	47.8	129.5	50.5	...	...	131.2	51.8	129.9 (123.8)	50.8 (46.1)	127.2	48.7

Table 5.3: Comparison of the force constants f (in Nm^{-1}) of the O–H$\cdots$O stretching mode ν (in cm^{-1}) of the four carboxylic acids. Anharmonic results of ν and f at the B3LYP/6-311++G(2d,2p) level are listed in parentheses.

One can thus conclude that the hydrogen bonds are stiffer in acetic acid than in formic acid, but the dominant reason for this is seen to be a nuclear quantum effect related to the librational zero point motion in the light formic acid unit, as discussed before [3]. It weakens the hydrogen bonds by spreading the hydrogen locations away from the optimum hydrogen bond geometry, even at 0 K. This off-diagonal quantum effect is captured by the perturbative treatment. It is largely attenuated in the heavier carboxylic acid units, making their hydrogen bonds effectively stiffer than those in formic acid dimer. This trend is overlaid by an increasing mismatch between hydrogen bond donor and acceptor strengths, if larger substituents are introduced. In the case of propiolic acid, mesomeric effects are likely to contribute. It may be rewarding to study these effects at the molecular orbital level, whereas here we concentrate on the observable force constant trends.

The next question to be answered is whether this local hydrogen bond stiffness trend is reflected in global binding energies. The latter are difficult to access experimentally in the 0 K limit [19] and the room temperature values [140] of $\approx 62 \, kJ/mol$ for formic and acetic acid dimers suffer from thermal distortions and experimental uncertainty. However, their prediction at the various quantum chemical levels of approximation can be discussed and is summarized in Tab. 5.4 and shown in Fig. 5.3. The electronic binding energy, which is free of zero point motion effects, is indeed minimal for formic acid dimer (dashed lines). It is higher for acetic acid dimer at all levels, although by varying amounts (3-6%). Whether or not it increases for pivalic and to a lesser degree propiolic acid depends on

Dimer	MP2				B3LYP				B97D	
	6-31+G*		6-311+G*		6-31+G*		6-311++G(2d,2p)		TZVP	
	D_0	D_e	D_0	D_e	D_0	D_e	D_0	D_e	D_0	D_e
$(HCOOH)_2$	58.3	67.6	52.8	60.3	57.9	66.2	55.5	63.5	60.5	67.3
$(CH_3COOH)_2$	63.3	70.5	59.1	64.0	61.6	68.3	59.1	65.2	65.4	70.2
$(HCCCOOH)_2$	61.8	69.7	57.7	62.3	60.9	67.1	58.2	63.5	63.4	67.8
$(C(CH_3)_3COOH)_2$	68.4	74.8	...	66.6	62.8	68.3	60.1	65.7	67.9	71.5

Table 5.4: Comparison of D_e and D_0 (harmonic) in $kJ \, mol^{-1}$ for the dimers of the four carboxylic acids.

the capture of dispersion energy in the theoretical treatment, despite the large distance between the two R-groups in the dimer. At dispersion-corrected B97D level and at MP2 level for large basis sets, the dissociation energy clearly increases. At B3LYP level using a sufficiently large basis set, there is almost no change. The electronic structure trends are typically amplified upon inclusion of harmonic zero point energy correction (full lines). Now, the instability of formic acid dimer relative to acetic acid dimer is doubled (6-12%), because its hydrogen bonds are weakened most by zero point motion. We refrain from providing anharmonic values for D_0, because the perturbative approach fails for large amplitude torsional modes. Although this work is more concerned with relative values, we note that the harmonic B97D value for acetic acid dimer of $D_0 = 65\,\text{kJ/mol}$ is in good agreement with a recent analysis and extrapolation of experimental data [19] (65-66 kJ/mol).

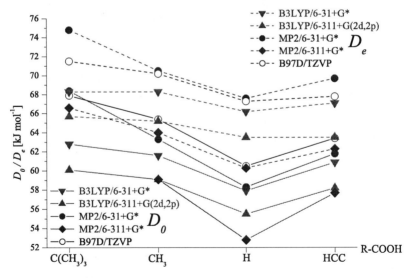

Figure 5.3: Comparison of calculated dimer dissociation energies D_0 (solid lines)/D_e (dashed lines) of the four carboxylic acids (in kJ mol^{-1}).

Taking acetic acid as a reference, one may conclude that formic acid dimer is more weakly bound because of its zero point amplitude and because of the lack of an electron donating effect into the carboxyl group. The former effect clearly dominates and reduces the stiffness of the corresponding hydrogen bonds. If the methyl groups are exchanged by *t*-butyl groups, the dimer dissociation energy increases, presumably due to dispersion forces between the remote groups. However, this gain in cohesion is not reflected by the stretching force constant. The latter drops, possibly as a consequence of mismatch in donating and accepting power of the carboxylic acid dimer unit.

Apart from the subtle quantum dynamical effect in formic acid dimer, the best estimates for hydrogen bond stiffness and dimer dissociation energy are thus seen to anti-correlate. The stiffest hydrogen bonds give rise to the most weakly bound dimers. This may serve as a precaution against correlating local hydrogen bond strength to global cohesion energy in dimers, as emphasized before [47].

5.1.3 Conclusions

A pseudo-diatomic analysis of the hydrogen bond stretching modes in four carboxylic acid dimers showed that acetic acid features the stiffest hydrogen bonds, largely due to reduced librational amplitude compared to formic acid dimer, which also leads to a larger dissociation energy of acetic acid dimer.

Otherwise, the stiffness and dissociation energy of carboxylic acid dimers appear to be qualitatively anti-correlated, underscoring the difference between local and global interactions in hydrogen-bonded dimers.

5.2 Propiolic and pivalic acid: region above 275 cm^{-1}

5.2.1 Propiolic acid

Propiolic acid monomer has 15 vibrational fundamentals and the dimer has 36 modes. The monomer fundamentals and Raman active dimer fundamentals under 275 cm^{-1} have been discussed (see Fig. 5.1 and Tab. 5.1). All the remaining fundamentals of the monomer as well as the remaining Raman active fundamentals of the dimer are observed in this work. Comparisons of GP and jet spectra are shown in Figs. 5.4 and 5.5. All the spectra are combined with four measurements in different wavenumber regions under equivalent conditions. The 67 L stainless steel reservoir and a homebuilt 8.0×0.05 mm^2 slit nozzle were used in the jet measurements. The chamber was evacuated only by a 250 m^3/h Roots pump backed by a 100 m^3/h rotary vane pump.

Band positions and assignments are listed in Tab. 5.5 and compared with the quantum chemical calculations. The good agreement of the experimental band positions with those from calculations, especially from the anharmonic calculations, makes the assignment straightforward.

Complicated band structures for the δ(O–H), ν(C=O) and ν(C≡C) modes are observed (see Figs. 5.4 and 5.5). The two bands marked with * near to the intense C–C stretching band in Fig. 5.4 could be due to the ^{13}C isotopomer, as in the case of AAD. For both monomer and dimer, the weak band is about 15 cm^{-1} red shifted from the main band (803 cm^{-1} for the * band in the GP spectrum and 849 cm^{-1} for that in the jet spectrum). The intensity ratio of the band marked with * and the related ν(C–C) main band is around 4%. A similar isotope effect is assumed for the CC triple bond but not verified due to its complicated band structure (see Fig. 5.5).

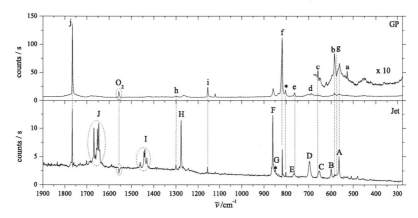

Figure 5.4: Comparison of the Raman spectra of propiolic acid between 275 and $1900\,cm^{-1}$. The jet spectrum is a combination of four measurements under the same conditions of trace c) in Fig. 5.1. The GP spectrum is a combination of four measurements under the following conditions: $T_s = 16°C$ ($< 0.5\%$ HCCCOOH in He), $p_b = 30\,mbar$, $6 \times 30\,s$.

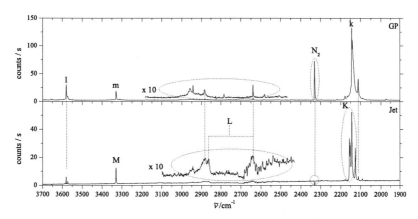

Figure 5.5: Comparison of the Raman spectra of propiolic acid between 1900 and $3700\,cm^{-1}$. Measurements conditions are the same as those in Fig. 5.4.

Mode	Dimer				Monomer		
	Label	Sym.	Exp.	Cal.	Label	Exp.	Cal.
δ(OCO)	A	A_g	565	569.2(568.7)	a	528	528.4(527.1)
δ(CCC)	B	A_g	600	608.9(602.5)	b	586	596.8(591.4)
δ(C–H)	C	A_g	654	683.7(670.5)	c	658	682.8(667.9)
γ(C–H)	D	B_g	697	738.6(725.3)	d	688	734.7(712.2)
γ(OCO)	E	B_g	768	779.4(767.6)	e	763	774.4(764.7)
ν(C–C)	F	A_g	861	868.7(861.1)	f	818	822.0(810.4)
γ(O–H)	G	B_g	857	967.8(859.3)	g	577	593.1(576.3)
ν(C–O)	H	A_g	1277	1295.4(1261.4)	h	1300	1358.0(1292.8)
δ(O–H)	I	A_g	1440	1492.9(1457.8)	i	1154	1167.3(1127.4)
ν(C=O)	J	A_g	1647	1679.5(1632.0)	j	1766	1779.6(1750.1)
ν(C≡C)	K	A_g	2144	2220.8(2185.2)	k	2144	2218.2(2185.8)
ν(O–H)	L	A_g	~2750[a]	3077.9(2703.0)	l	3580	3753.3(3553.4)
ν(C–H)	M	A_g	3330	3461.3(3328.4)	m	3330	3461.2(3335.5)

[a] Average value of the two assumed Fermi resonance bands, see Fig. 5.5.

Table 5.5: Band positions (in cm^{-1}) and assignments of the propiolic acid monomer fundamentals and the Raman active fundamentals of dimer above 275 cm^{-1}, compared with the harmonic/anharmonic calculations (listed in parentheses) at the B3LYP/6-311++G(2d,2p) level (in cm^{-1}).

5.2.2 Pivalic acid

Cyclic pivalic acid dimer has in total 90 vibrational fundamentals, half of which are Raman active. Beside the 20 C/O–H stretching modes all the other vibrations are below 1800 cm^{-1}. The Raman vibrational spectra of the pivalic acid dimer and its deuterium isotopomers in the entire fundamental wavenumber region have been experimentally studied in the liquid [141] and solid [142] states and theoretically analyzed [143]. The IR jet spectrum was also reported [42].

The Raman active fundamentals of pivalic acid dimer below 275 cm^{-1} have been discussed (see Fig. 5.1 and Tab. 5.1). Fig. 5.6 shows its Raman jet spectrum between 250 cm^{-1} and 1750 cm^{-1}. The spectrum is a combination of three measurements in different wavenumber regions under equivalent conditions. Band positions and assignments are listed in Tab. 5.6, together with the former studies [42, 142], and compared with the quantum chemical calculations. Most bands show a good agreement with the calculated wavenumbers. Assignments of several bands are uncertain due to closely spared bands (e.g., bands J and K). Calculations of the pivalic acid monomer at the same level indicate several strongly Raman active bands at 569.2 cm^{-1}, 722.1 cm^{-1}, 864.5 cm^{-1}, which fit the experimental values (bands marked with * in Fig. 5.6 at 568 cm^{-1}, 715/723 cm^{-1} and 862 cm^{-1}) very well. Due to the overlap with dimer bands no further monomer bands can be distinguished in Fig. 5.6.

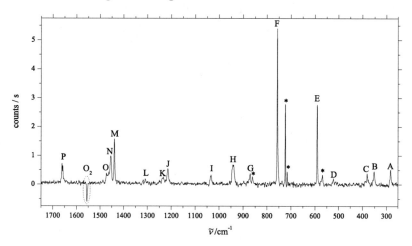

Figure 5.6: Raman jet spectra of pivalic acid dimer between 250 and 1750 cm^{-1}. $T_s = 16°C$ (~0.04% C(CH$_3$)$_3$COOH in He), $p_s = 1000$ mbar, $d = 1$ mm, 12×300 s, 4.0×0.15 mm^2 nozzle, 67 L reservoir, 250 m^3/h Roots pump and 100 m^3/h rotary vane pump. * indicates monomer signals.

Mode	Raman active			IR active	
	Jet[a]	Solid[b]	Cal.	Jet[c]	Cal.
δ(CCC)	282(A)		280.5	...	291.1
τ(CH$_3$)	282(A)	287	284.1	...	282.1
γ(CCC)	352(B)	359	350.6	...	374.3
ν(O–H...O) + SD(CCC)	382(C)	384	387.0	...	357.6
DD(CCC)	...	...	380.8	...	379.1
δ(O–H...O) + δ(CCC)	522(D)	524	521.2	...	547.5
δ(OCO) + + SD(CCC)	591(E)	594	590.5	...	587.4
δ(OCO)	757(F)	758	755.3	...	770.1
γ(OCO)	...	...	776.9	...	772.6
ν(C–C)	871(G)	872	877.3	...	879.2
Ro(CH$_3$)	942(H)	943	946.9	...	947.0
Ro(CH$_3$)	...	...	953.4	...	953.4
δ(O–H)	...	...	965.5	...	1008.0
Ro(CH$_3$)	...	...	979.8	...	979.6
Ro(CH$_3$)	1034(I)	1033	1057.0	...	1057.3
γ(CCO) + Ro(CH$_3$)	1034(I)	1033	1058.6	...	1058.6
δ(OCO) + Ro(CH$_3$)	1215(J)	1212	1230.4	1210/1219	1232.4
γ(CCC) + Ro(CH$_3$)	1236(K)	1232	1235.7	1232	1236.2
δ(CCC) + Ro(CH$_3$)	...	...	1255.2	...	1254.4
ν(C–O)	1309(L)	1297	1329.6	1319	1338.4
DD(CH$_3$)	...	...	1418.0	1423	1418.0
DD(CH$_3$) + δ(O–H)	1439(M)	1431	1422.5	1423	1419.1
DD(CH$_3$) + δ(O–H)	1454(N)	1453	1450.1	...	1437.0
SD(CH$_3$) + δ(O–H)	1473(O)	1463	1474.2		1462.9
DD(CH$_3$)	...	1483	1501.7	1489	1501.8
DD(CH$_3$)	...	...	1509.4	...	1509.2
DD(CH$_3$)	...	...	1513.6	...	1513.8
DD(CH$_3$)	...	...	1524.0	...	1524.0
SD(CH$_3$)	...	...	1526.9	1539	1526.6
SD(CH$_3$)	...	...	1544.8	1558	1545.8
ν(C=O)	1658/1661(P)	1657	1708.4	1718	1752.3

[a] Raman jet spectra, this work. See Fig. 5.6.

[b] Longueville *et al.*, Raman spectra in solid state, Ref.142.

[c] Emmeluth, FTIR jet spectra, Ref.42.

Table 5.6: Comparison of the experimental band positions of the fundamentals of pivalic acid dimer between $250\,\mathrm{cm}^{-1}$ and $1750\,\mathrm{cm}^{-1}$ with the harmonic quantum chemical calculations at the B3LYP/6-31+G(d) level (in cm^{-1}). Dimer stretching vibrations are marked ν, in-plane bending modes δ, out-of-plane bending modes γ, torsion modes τ, degenerate rocking Ro, degenerate deformation DD and symmetric deformation SD.

6 Water clusters

Water plays a crucial role in living organisms and it is considered as one of the key factors for the success of life on earth. This small molecule is the most investigated and intriguing molecule of our world and the number of scientific articles, monographs and textbooks focusing on it is enormous, surpassing the order of magnitude of 10^5 [144]. In recent years, many experimental tools have provided a rich but still incomplete picture of the effects of water on structure, dynamics, and activity of biomolecular systems. Vibrations of the water monomer have been well understood by high resolution IR and Raman studies [145–150]. Beside this, the exploration of the structural and binding properties of water clusters provides a key for the understanding of many natural phenomena in biological and chemical systems with hydrogen bonds.

Isolated water clusters can only be formed at very low temperature. Different from the carboxylic acids, which have a strong preference for cyclic dimers, water will always strive to continue aggregation to form larger clusters, unless they are kinetically hindered to do so. There are many cluster bands observed in the IR and Raman spectra, but the band positions and intensities often have no clear systematics. The correct assignment of every band of the water clusters is a substantial challenge.

Numerous experimental [1, 12, 151–166] and theoretical studies [167–175] on water clusters have been published, focused on the detailed characterization of the small water complexes $(H_2O)_n$ ($n = 2$-6). The theoretical work of Xantheas has confirmed that the energetic global minimum of small water clusters $((H_2O)_n$, $n = 3$-5) is cyclic [171] and this high symmetry of the cyclic clusters makes IR and Raman techniques rather complementary.

IR absorption spectroscopic studies have already been carried out with different methods in different chemical environments, such as infrared cavity ring-down laser absorption spectroscopy (IR CRDLAS) [159, 160], IR predissociation spectroscopy [151, 152], size-selective IR-depletion scattering experiments [1] and direct absorption FTIR spectroscopy [164]. Compared with IR studies the Raman measurement of water clusters is technically demanding because the inelastic scattering power of water clusters is very low. CARS spectroscopy in supersonic jet was used to study water clusters and the O–H stretching vibrations of water dimer and trimer were assigned [154]. Bands at $3480\,cm^{-1}$, $3366\,cm^{-1}$ and $3340\,cm^{-1}$ were observed but the assignment exhibited poor compliance with quantum mechanical predictions and was contradicted by Nelander with respect to GP and Matrix isolation measurements [156].

Almost all the theoretical studies are based on the IR experimental studies. The lack of reliable Raman experimental data has slowed down theoretical analysis and made a clear assignment of the IR bands difficult. In the present work, the hydrogen-bonded O–H stretching vibrations of the neutral water clusters from dimer up to hexamer were studied using the *curry*-jet with different carrier gases. Heavy water (D_2O) was investigated under the same conditions for comparison. Assisted by the former IR jet study in our group [164] and quantum chemical calculations, several Raman active water cluster bands can be assigned, with evidence for a single (the most stable) isomer for each cluster size (up to the pentamer). More information comes from relaxation experiments by using different distances from the nozzle exit and by using different carrier gases. Many sub-peaks from Davydov splittings, hot bands and tunneling splittings are expected.

6.1 Band assignment

For all the Raman spectra in this chapter, the 67 L stainless steel reservoir and the $8.0 \times 0.05\,mm^2$ slit nozzle were used. The chamber was evacuated by a $250\,m^3/h$ Roots pump backed by a $100\,m^3/h$ rotary vane pump. Normally the stagnation pressure p_s was set to higher than 1 bar and the nozzle distance larger than $2\,mm$

to improve the extent of aggregation. The relatively large nozzle distances reduce the weak Raman signal intensities of the water cluster bands further and enhance the difficulty of the band assignment. It should be noted that the pressure of the carrier gas in the saturator p_0 will be higher than p_s. When p_s exceeds 1 bar, the assumption of Eq. 2.2 ($p_0 = 1$ bar) is no more valid. In these cases, $p_0 = 1.5$ bar is used to estimate the "maximum" H_2O/D_2O concentrations.

Jet spectra of the hydrogen-bonded O–H stretching vibrations of the H_2O clusters between 3170 and 3630 cm^{-1} under different measurement conditions are shown in Fig. 6.1. Generally, the relative intensities of the bands from larger

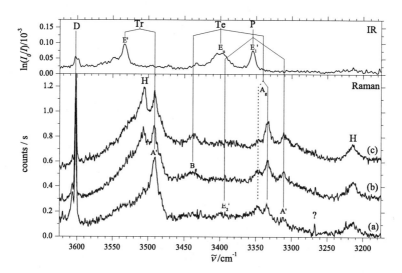

Figure 6.1: Jet spectra of the hydrogen-bonded O–H stretching vibrations of H_2O clusters. Top trace: Ragout-jet FTIR spectrum of water in He, taken from Ref. 164; bottom traces: Raman relaxation study with different carrier gases: (a) pure He, (b) He/Ne $\approx$ 5:1, (c) He/Ar $\approx$ 5:1. All Raman spectra were recorded under the following conditions: $T_s = 22$°C (1.7% H_2O), $p_s = 1200$ mbar, $d = 2$ mm, 12 × 300 s. The sharp peak at 3267 cm^{-1} (marked with "?") could be the $2\nu_2$ (bending) band of the dimer, because a strong and sharp fundamental bending band of dimer was observed at 1629 cm^{-1} in the IR cavity ringdown spectroscopy [176].

clusters increase faster than those from smaller clusters with better cooling by a more effective carrier gas (from He to He + Ne, then He + Ar).

The following symbols are used for the assignment of the water clusters: D for dimer, Tr for trimer, Te for tetramer, P for pentamer and H for hexamer. The largest cluster observed in the IR spectrum in Fig. 6.1 is the pentamer. The strong cyclic hexamer band at $3335 \, \text{cm}^{-1}$ in liquid He [172] is not observed in the IR jet spectrum in Fig. 6.1 because it dose not correspond to the global minimum structure. Therefore, quantum chemical calculations in this work were carried out up to cyclic pentamer. Only the bands from dimer to pentamer are relatively safely assigned. Symmetries of the two Raman active bands believed to come from the cage hexamer are not given.

Symmetries of other bands are given in Fig. 6.1 based on the assumption that each of the cyclic clusters has a dynamically averaged planar structure with C_{nh} symmetry ($C_{3h}/C_{4h}/C_{5h}$ for trimer/tetramer/pentamer, see Fig. 6.2). This assumption is also used for the quantum chemical calculations. There is more than one calculated wavenumber for the same band in the literature because the wavenumbers of the clusters are usually calculated in the "real" non-planar geometries. Normally in these cases, the wavenumber differences between the sub bands are rather small (within $10 \, \text{cm}^{-1}$, see Tab. 6.3 later on) and the average values of the band positions are used for the following analyses.

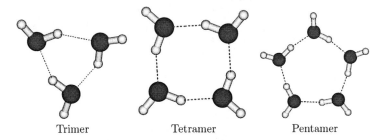

| Trimer | Tetramer | Pentamer |

Figure 6.2: Planar structures of the cyclic water clusters $((H_2O)_n, \ n = 3\text{-}5)$ with C_{nh} symmetry, all optimized at the B3LYP/6-31+G* level. Only the hydrogen-bonded O–H stretching vibrations are discussed in this work.

Wavenumbers and assignments of the bands observed in Fig. 6.1 are listed in Tab. 6.1, compared with some former (IR) studies. The several IR active bands have been relatively safely assigned [164] and a good agreement is seen between the band positions from different studies (see Tab. 6.1). Generally, all the bands observed in the He droplet measurements are a little red shifted compared to those in the jet experiments, maybe due to the lower vibrational temperatures. Therefore, these IR bands will be used as "benchmark" for the assignment of the more complicated Raman active bands.

		Jet			Liquid He	Matrix isolation	
n	symmetry	Göttingen	IR^a	IR^b	IR^c	IR^d	IR^e
2	A'	3602^f	3601	3601, 3572^g	3597	3629	
2	A'	3602^h					
3	E'	3548^f (54)			3544		
		3533^f (69)	3533	3532, 3517^g	3529	3507	3530
6		3506^h (96)					
3	A'	3491^h (111)					
4	B_g	3438^h (164)					
5	E'_2	3410^h (192)					
4	E_u	3401^f (201)	3416	3399, 3372^g	3394		3382
5	E'_1	3355^f (247)	3360	3355, 3330^g	3353	3368	
4	A_g	3347^h (255)					
		3334^h (268)					
5	A'	3311^h (291)					
6		3214^h (388)					

[a] Huisken et al., size-selected IR jet spectrum, Ref. 158
[b] Moudens et al., IR jet spectrum, Ref. 177
[c] Burnham et al., liquid helium droplett IR spectrum, Ref. 172
[d] Ohno et al., IR spectrum in $O_2 + N_2$ matrix, Ref. 165
[e] Ceponkus et al., IR spectrum in Ne matrix, Ref. 166
[f] Nesbitt et al., IR jet spectra, Ref. 164
[g] Ar solvated clusters.
[h] Raman jet spectra, this work.

Table 6.1: Band positions (in cm^{-1}) and assignments of the hydrogen-bonded O–H stretching vibrations of H_2O cluster bands in Fig. 6.1, compared with some former studies. The wavenumber shifts (in cm^{-1}) between the dimer band and other cluster bands observed in the jet experiments are listed in parentheses.

Jet spectra of the D_2O clusters between 2350 and 2650 cm^{-1} are shown in Fig. 6.3. The band shapes of the D_2O clusters are very similar to those of the H_2O clusters under the same measurement conditions, although the band intensities are significant higher (see Fig. 6.4). Therefore, the wavenumber sequence of the bands from different clusters should be the same for H_2O and D_2O clusters, although due to the larger deuterium mass the wavenumber difference between the cluster bands of D_2O is smaller than the related one of the H_2O clusters. Wavenumbers and assignments of the cluster bands are listed in Tab 6.2. As in the H_2O clusters, the IR active bands of the D_2O clusters have been assigned [164] and will be used as "benchmark" for the following assignment as well.

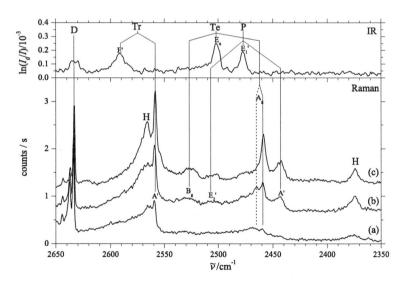

Figure 6.3: Jet spectra of the hydrogen-bonded O–D stretching vibrations of D_2O clusters. Top trace: Ragout-jet FTIR spectrum of D_2O in He, taken from Ref. 164; bottom traces: Raman relaxation study with different carrier gases: (a) pure He, (b) He/Ne ≈ 5:1, (c) He/Ar ≈ 5:1. All Raman spectra were carried out under the following conditions: $T_s = 22°C$ (1.7% D_2O), $p_s = 1100$ mbar, $d = 2$ mm, 12×300 s.

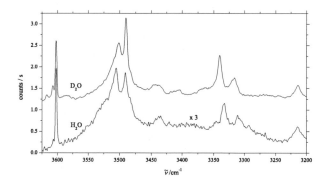

Figure 6.4: Comparison of the Raman jet spectra of H_2O and D_2O clusters scaled to similar scattering intensity. The bottom trace is the same as trace (c) from Fig. 6.1 and the top trace is the same as trace (c) from Fig. 6.3. Wavenumbers of D_2O clusters are scaled by 1.498 and then $342.4\,\text{cm}^{-1}$ is subtracted.

n	symmetry	Jet	Gas phase[a]	Matrix isolation
2	A'	2633^b		2613.9^c
2	A'	2633^d		
3	E'	2591^b (42)	2588	2584.7^e
3	A'	2566^d (67)		
6		2559^d (74)		
4	B_g	2527^d (106)		
5	E_2'	2507^d (126)		
4	E_u	2502^b (131)		2494.1^e
5	E_1'	2477^b (156)		
4	A_g	2465^d (168)		
		2459^d (174)		
5	A'	2443^d (190)		
6		2374^d (259)		

[a] Paul *et al.*, cavity ring-down IR spectrum, Ref. 160.
[b] Nesbitt *et al.*, IR jet spectra, Ref. 164
[c] Engdahl and Nelander, IR spectrum in Ar matrix, Ref. 12.
[d] Raman jet spectra, this work.
[e] Ceponkus *et al.*, IR spectrum in Ne matrix, Ref. 166.

Table 6.2: Band positions (in cm^{-1}) and assignments of the hydrogen-bonded O–D stretching vibrations of D_2O cluster bands in Fig. 6.3, compared with some former studies. The wavenumber shifts (in cm^{-1}) between the dimer band and other cluster bands observed in the jet experiments are listed in parentheses.

With these relatively certain "benchmarks", the quality of the quantum chemical calculations using different methods and levels is firstly tested. The calculated band positions of the planar $(H_2O)_n$ clusters $(n = 2\text{-}5)$ are shown in Tab. 6.3. Generally, the calculated "absolute" band positions do not fit well to the experimental values. Nevertheless, almost all the calculations show the same wavenumber sequence of the cluster bands which is very helpful for the assignment of the Raman active bands: $D > Tr\ (E') > Tr\ (A') > Te\ (B_g) > P\ (E'_2) \approx Te\ (E_u) > P\ (E'_1) > Te$ $(A_g) > P\ (A')$. This sequence is also applied to the assignment of the D_2O clusters.

The calculation with more expensive methods and large basis sets appear to fit better to our assignment. Comparisons of experimental and calculated wavenumber shifts between the bands of $(H_2O)_n$ $(n = 3\text{-}5)$ and water dimer are shown in Fig. 6.5. The calculations at the MP2/aug-cc-pVDZ level show the best agreement with the experimental results. All the calculated wavenumber shifts between the

		MP2		HF	B3LYP	
n	Symmetry	aug-cc-pVDZ[a]	6-31+G*[b]	6-31+G*[c]	6-31+G*[b]	6-311+G(d,p)[e]
2	A'	3704	3703	3597	3670	3644
3	E'	3641 (63)	3657 (46)	3562 (35)	3618 (52)	3507 (137)
		3632 (72)		3558 (39)		3499 (145)
3	A'	3573 (131)	3623 (80)	3519 (78)	3583 (87)	3455 (189)
4	B_g	3522 (182)	3607 (96)	3521 (76)	3548 (122)	3416 (228)
5	E'_2	3494 (210)	3578 (125)	3513 (84)	3504 (166)	3391 (253)
		3487 (217)		3509 (88)		3384 (260)
4	E_u	3484 (220)	3585 (118)	3496 (101)	3525 (145)	3388 (256)
		3484 (220)		3496 (101)		3387 (257)
5	E'_1	3442 (262)	3546 (157)	3479 (118)	3468 (202)	3349 (295)
		3433 (271)		3476 (121)		3343 (301)
4	A_g	3391 (313)	3534 (169)	3442 (155)	3466 (204)	3315 (329)
5	A'	3354 (350)	3495 (208)	3431 (166)	3406 (264)	3277 (367)

[a] From Ref. 172, clusters with "real" non-planar geometries.
[b] This work, clusters with planar C_{nh} symmetries.
[c] From Ref. 173, clusters with "real" non-planar geometries.
[d] From Ref. 165, clusters with "real" non-planar geometries.

Table 6.3: Comparison of the calculated wavenumbers (in cm^{-1}) of the hydrogen-bonded O–H stretching vibrations of the water cyclic clusters $(H_2O)_n$ $(n = 2\text{-}5)$ using different quantum chemical methods and basis sets. The calculated wavenumber shifts (in cm^{-1}) between the dimer band and other cluster bands are listed in parentheses.

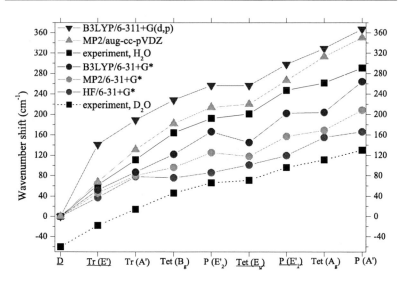

Figure 6.5: Comparison of experimental and calculated wavenumber shifts between the hydrogen-bonded O–H stretching vibrational bands of $(H_2O)_n$ ($n = 3$-5) and that of water dimer. The wavenumber shifts are listed in Tabs. 6.1 (exp.) and 6.3 (cal.). Average experimental/theoretical values are used for the symmetries with more than one sub band. The related wavenumber shifts of D_2O have also been shown for comparison with a dotted line and were shifted down by $60 \, cm^{-1}$ for clarity

bands of $(H_2O)_n$ ($n = 3$-5) and water dimer at this level are systematically overestimated by about 10%.

Assisted by this wavenumber sequence the tentative assignment shown in Tabs. 6.1 and 6.2 can be made. They can be tested with the relaxation experiments and a depolarization analysis. Comparisons of the Raman spectra of H_2O/D_2O clusters using different carrier gases are shown in Figs. 6.1 and 6.3 and have been discussed. The cooling effect on the relative band intensities can be controlled by using different nozzle distances (see Fig. 6.6). A similar thermal effect is observed.

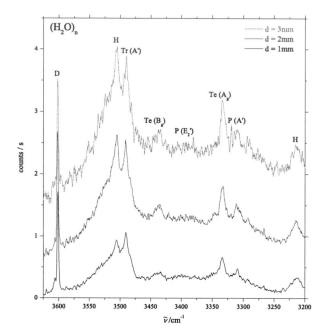

Figure 6.6: Comparison of the Raman spectra of the hydrogen-bonded O–H stretching vibrations of H$_2$O clusters measured at different nozzle distances d, scaled to similar scattering intensity of the dimer band. $T_s = 22°C$ (1.7% H$_2$O in He/Ar $\approx$ 5:1), $p_s = 1100$ mbar, 12 × 00 s.

Raman depolarization analysis (Fig. 6.7) shows that only the two bands at 3438 cm^{-1} and 3410 cm^{-1} in the spectra of the H$_2$O clusters may be totally depolarized ($\rho_\perp = 0.75$). According to the calculations the B$_g$ band of tetramer and E$'_2$ band of the pentamer have the depolarization ratio of 0.75 (see Tab. 6.4). It fits our assignment quite well. These two bands are relatively weak in the spectra, also in agreement with the theoretical prediction: Their Raman activities are much smaller than those of the other Raman active bands (the E$'$ band of the water trimer is both IR and Raman active). The same applies to the two related bands of the D$_2$O clusters at 2527 cm^{-1} and 2507 cm^{-1} assigned to the same sym-

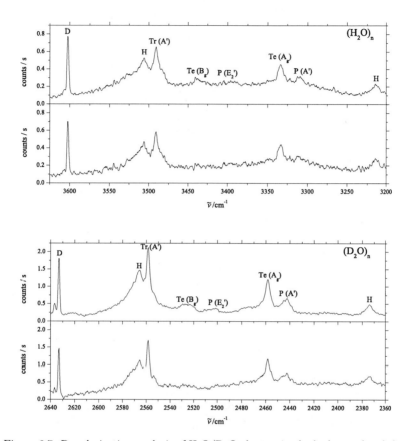

Figure 6.7: Depolarization analysis of H_2O/D_2O clusters in the hydrogen-bonded O–H/D stretching vibrational region, with the excitation laser perpendicular to the scattering plane (top trace) and residual after subtracting 7/6 of the spectrum obtained with the excitation laser parallel to the scattering plane. All the spectra were recorded under the following conditions: $T_s = 22°C$ (1.7% H_2O/D_2O in He/Ar ≈ 5:1), $p_s = 1100$ mbar, $d = 2$ mm, 12×300 s.

n	symmetry	MP2/6-31+G*	B3LYP/6-31+G*
2	A′	140.8 (0.13)	157.4 (0.17)
3	E′	20.1 (0.75)	29.9 (0.75)
		20.1 (0.75)	29.9 (0.75)
3	A′	252.6 (0.08)	268.8 (0.09)
4	B_g	75.1 (0.75)	107.5 (0.75)
5	E_2'	57.7 (0.75)	81.0 (0.75)
		57.7 (0.75)	81.0 (0.75)
4	E_u	0	0
		0	0
5	E_1'	0	0
		0	0
4	A_g	365.5 (0.08)	387.8 (0.09)
5	A′	488.1 (0.08)	525.5 (0.10)

Table 6.4: Comparison of the calculated Raman activities of the hydrogen-bonded O–H stretching vibrations of the planar cyclic water clusters $(H_2O)_n$ ($n = 2$-5) using different quantum chemical methods. The calculated depolarization ratio $\rho_\perp$ are listed in parentheses.

metries. Due to the higher band intensities of the D_2O clusters it is even easier to determine the disappearance of the two bands.

It should be noted that in some cases more than one peak is observed in the spectra for the same band. For example, the weak band at $3548\,cm^{-1}$ in the IR spectra was observed at $3544\,cm^{-1}$ in the liquid helium droplet experiment [172] with no assignment. A corresponding shift of $15\,cm^{-1}$ between this band and the strong trimer E′ main band is found in the jet IR experiment [164]. Therefore, it may come from the tunneling splittings of the E′ band of the trimer. The average value of these two bands is used as the band position of the Tr (E′) band for the analysis in this work (e.g., in Fig. 6.5). This effect was not observed in the IR spectrum of the D_2O clusters. This is not surprising because the wavenumber difference between these two related bands of D_2O trimer should be much smaller and may be below the spectral resolution of the IR spectrometer used in Fig. 6.3.

A clear tunneling splitting character was observed for the Raman active A_g band of the H_2O/D_2O tetramer. The two sub bands have similar intensities in the measurement using pure helium as carrier gas. With better cooling using more effective carrier gases, the relative intensity of the band with lower wavenumber is always increasing compared to that of the band with higher wavenumber (see Figs. 6.1 and 6.3). The former one should be energetically more stable. The splitting of the water bands is $13\,cm^{-1}$, compared the $6\,cm^{-1}$ of D_2O tetramer. In contrast, in the case of the Raman active A' band of the trimer, the higher wavenumber component is almost not observed in the spectrum using pure helium as carrier gas but its relative intensity increases with better cooling, almost synchronous to the band with the lowest wavenumber in the Raman spectra. Besides, no correspondence of the intensity ratio of this band and the A' trimer band is found for H_2O and D_2O under similar measurement conditions. Therefore, this band is believed to come from the cage hexamer, like the band with the lowest wavenumber in the Raman spectra. The latter one was observed at $3214\,cm^{-1}$ in our Raman jet spectra, similar to the cage hexamer band observed at $3220\,cm^{-1}$ in the former GP IR spectra [159] as well as the band at $3229\,cm^{-1}$ in the liquid helium droplet IR spectrum [172].

When comparing the jet data to the matrix isolation spectra, there are always relatively large wavenumber shifts between them for the same bands. It would be helpful if we could observe these shifts stepwise in the jet measurements using neon/argon as the carrier gas. However, the flow of pure Ne/Ar prevents sufficiently high stagnation pressures p_s at the water concentration needed to overcome the weak Raman scattering strength of the water clusters. No clear cluster bands can be observed in the jet spectra with pure Ne/Ar as the carrier gas. Fig. 6.8 shows a measurement series under similar conditions but with different mole ratios of D_2O and Ar in the gas mixture. At first, pure helium picked up a small fraction of D_2O vapor through a saturator, was mixed with the pure argon through another saturator and went into the reservoir. The pressure of the helium was set to 1.5 bar. With its help p_s could be set to 1.2 bar. With the different pressures of argon and the various vapor pressures of D_2O controlled by the saturator temperatures, different mole ratios of D_2O and Ar were obtained and the Ar nanocoating effect on the symmetrical O–D stretching vibrations of the D_2O clusters could be

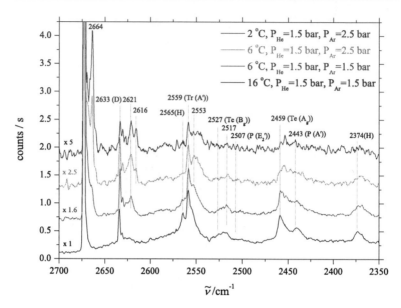

Figure 6.8: Ar nanocoating effect on the hydrogen-bonded O–D stretching vibrations of the D_2O clusters scaled to similar scattering intensity. For all the spectra, $p_s = 1200$ mbar, $d = 2$ mm, 12×300 s. See text for details.

observed. Although the accurate concentration of D_2O were not known, there is a tendency of less D_2O in more argon from bottom to top in Fig. 6.8. Generally, the cluster bands on the top trace show a red shift of 6-12 cm^{-1} compared to those on the bottom trace. The new (dimer) band at 2616 cm^{-1} corresponds to the band observed at 2614 cm^{-1} in the Ar matrix experiment [12] (See Tab. 6.2). The band intensities of the two hexamer bands at 2565 cm^{-1} (shifted to 2553 cm^{-1}?) and 2374 cm^{-1} increase with more argon in the carrier gas, compared with the band intensities of the trimer A$'$ band at 2559 cm^{-1} and the tetramer A$_g$ band at 2459 cm^{-1}. A red shift of 12 cm^{-1} was observed for the band at 2566 cm^{-1} and 6 cm^{-1} for the other hexamer band.

On the blue-shifted side of H_2O hexamer band at $3506\,\mathrm{cm}^{-1}$, a weak and broad band shoulder is observed (see Fig. 6.1). It should come from the E′ band of trimer because all the symmetries used in Fig. 6.1 lead to either IR or Raman activity except this one (see Tab. 6.4). The E′ band of trimer is both IR and Raman active. A similar effect is also observed for the related band of D_2O (see Fig. 6.3).

6.2 Davydov splitting analysis

According to the energy diagrams of the Davydov splittings of the symmetrical cyclic trimer and tetramer shown in Fig. 6.9, the wavenumber difference between the B_g band and the E_u band of the tetramer should be smaller than that between the E_u band and the A_g band. This precondition is fulfilled in our assignment, both for H_2O and D_2O. The Davydov splittings of the symmetrical O–H/D stretching

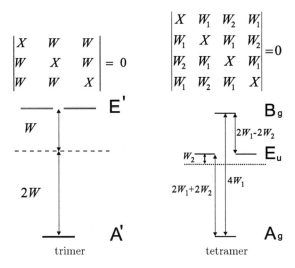

Figure 6.9: Energy diagrams of the Davydov splittings of symmetrical cyclic trimer and tetramer.

vibrations of the cyclic planar H_2O/D_2O trimer and tetramer are calculated and listed in Tab. 6.5. Average experimental wavenumbers of the bands assigned to the same symmetries are used.

The ratios of the several Davydov splittings of H_2O clusters relative to those of the D_2O clusters are similar: 1.54 for W of the trimer, 1.50 for W_1 and 1.57 for W_2 of the tetramer. The small differences are assumed to come from the errors of the experimental band centers due to the relatively weak and broad band shapes.

	$(H_2O)_n$	$(D_2O)_n$
$n = 3$		
E′	3540.5	2591
A′	3491	2559
W	16.5	10.7
$n = 4$		
B_g	3438	2527
E_u	3401	2502
A_g	3340.5	2462
W_1	24.4	16.3
W_2	11.8	7.5

Table 6.5: Davydov splittings of the hydrogen-bonded O–H/D stretching vibrations of the cyclic planar H_2O/D_2O trimer and tetramer (in cm^{-1}) from the experiment.

6.3 Conclusions and Outlook

The assignment of the symmetrical O–H/D stretching vibrations of the cyclic planar neutral H_2O/D_2O clusters is based on the quantum chemical calculations and the assumption of a single (the most stable) isomer for each cluster size (up to the pentamer). It is well confirmed by different measurement methods, including the depolarization analysis and relaxation experiments by using different distances from the nozzle exit and by using different carrier gases. Davydov splittings of the H_2O/D_2O trimer and tetramer and tunneling splittings of some bands are analyzed. The results may contribute to the construction of harmonic/anharmonic force fields of different water clusters.

The main difficulty on the Raman measurement is the weak, broad and overlapping bands of the water clusters. Relatively high stagnation pressures (generally higher than 1 bar) and high saturator temperatures (generally at 22°C) were used to improve the signal-noise ratio. However, to improve the extent of aggregation, nozzle distances beyond 2 mm were typically used. A Raman setup with heatable nozzle is currently being constructed in our lab. It may assist the water cluster measurements.

Pseudo size-selective methods with re-heated clusters during their deceleration in the receding Mach disk zone, e.g. a continuous spectral series with increasing background pressure in a jet expansion for a given nozzle distance and stagnation pressure, are limited. Besides, due to the rapid H/D exchange, measurements of mixed H_2O/D_2O with different mole ratios did not work.

7 Conclusions and Outlook

In the present work the carboxylic acid dimers and water clusters are chosen as model systems of strong and moderately strong hydrogen bonds and studied with spontaneous Raman spectroscopy coupled with a supersonic jet expansion. The results show that the *curry*-jet is a powerful tool to investigate hydrogen bonded systems.

The Raman spectra of formic acid and its three deuterium isotopomers are measured and assigned in the entire fundamental wavenumber region in Chap. 3. An overview of the fundamental modes of the monomer and dimer is achieved by comparing our experimental results with former experimental results as well as with quantum chemical calculations. Many overtone/combination bands are observed and assigned for the first time. The same applies to the acetic acid dimer in Chap. 4. Another main point in this chapter is the influence of experimental conditions on the equilibrium of monomer/dimer/larger clusters. Solid evidence is found for the existence of supra-dimer aggregates. The intermolecular modes of the different carboxylic acid dimers are systematically discussed in Chap. 5. The hydrogen bonded O–H/D stretching bands of the neutral water clusters (up to the pentamer) are tentatively assigned in Chap. 6.

Many combined comparisons are made in this work, for example, the intermolecular modes of the different carboxylic acid dimers. However, many more comparisons can be made but have not been made yet in this work. The most important contribution of this work is maybe to supply a relatively accurate but still basic data base which can be used for further investigations. Many interesting phenomena are observed but cannot be perfectly explained, yet. For example, why are the C=O stretching fundamentals of AAD much simpler than those of FAD?

Why are the O–D stretching vibration bands so regular for both the O-deuterated formic acid and the O-deuterated acetic acid dimers? These questions need to be answered in the future.

The biggest challenge in this work is still the assignment of the extremely complex spectra [91, 108] of carboxylic acid dimers above $2000 \, cm^{-1}$. Nevertheless, the new experimental results in this work have accomplished several important intermediate steps. The accurate band positions of the fundamentals are measured, especially the intermolecular modes whose role in characterizing the O–H/D stretching band as well as in providing efficient energy relaxation channels should be emphasized [17]. The tentative assignment of many combination bands indicates that binary combinations dominate.

Consideration is given to the possibility of the existence of the energetically less stable *cis*-FA monomer but it is not found in our experiment. The hetero formic acid-water complex is believed to contribute to some "*cis*-FA-monomer-like" bands in the Raman spectra and a future measurement series in the $600\text{-}1000 \, cm^{-1}$ window may reveal the true nature of the various bands. Another model system which will be tried in the near future is the formic acid-methyl formate complex.

As already mentioned above, the *curry*-jet is a powerful tool to investigate hydrogen bonded systems. However, some studies in this work are still limited by its current technical parameters. For example, the signal-noise ratio is unsatisfactory in the measurements of water clusters, especially in the depolarization analysis (see Fig. 6.7). This problem may be solved with the heatable nozzle which is currently under test in our laboratory. Beside this, a much better spectral resolution is required to resolve the small FAD tunneling splittings as a function of vibrational excitation [4].

A Bibliography

[1] Udo Buck and Friedrich Huisken. Infrared spectroscopy of size-selected water and methanol clusters. *Chem. Rev.*, 100:3863–3890, 2000.

[2] Yohann Scribano and Claude Leforestier. Contribution of water dimer absorption to the millimeter and far infrared atmospheric water continuum. *J. Chem. Phys.*, 126:234301, 2007.

[3] P. Zielke and M. A. Suhm. Raman jet spectroscopy of formic acid dimers: low frequency vibrational dynamics and beyond. *Phys. Chem. Chem. Phys.*, 9:4528–4534, 2007.

[4] Z. Xue and M. A. Suhm. Probing the stiffness of the simplest double hydrogen bond: The symmetric hydrogen bond modes of jet-cooled formic acid dimer. *J. Chem. Phys.*, 131:054301, 2009.

[5] T. N. Wassermann, D. Luckhaus, S. Coussan, and M. A. Suhm. Proton tunneling estimates for malonaldehyde vibrations from supersonic jet and matrix quenching experiments. *Phys. Chem. Chem. Phys.*, 8:2344–2348, 2006.

[6] Nils O. B. Lüttschwager, Tobias N. Wassermann, Stéphane Coussan, and Martin A. Suhm. Periodic bond breaking and making in the electronic ground state on a sub-picosecond timescale: OH bending spectroscopy of malonaldehyde in the frequency domain at low temperature. *Phys. Chem. Chem. Phys.*, 12:820–8207, 2010.

[7] Martin Quack and Martin A. Suhm. Quasiadiabatic channels and effective transition-state barriers for the disrotatory in-plane hydrogen-bond exchange motion in $(HF)_2$. *Chem. Phys. Lett.*, 183(3,4):187–194, 1991.

[8] Gerhard Herzberg. *Molecular Spectra and Molecular Structure. II. Infrared and Raman Spectra of Polyatomic Molecules.* D. van Nostrand, 1945.

[9] John E. Bertie and Kirk H. Michaelian. The Raman spectra of gaseous formic acid-h_2 and -d_2. *J. Chem. Phys.*, 76:886–894, 1982.

[10] Van Zandt Williams. Infra-red spectra of monomeric formic acid and its deuterated forms. II. Low frequency region (2200-800 cm^{-1}). *J. Chem. Phys.*, 15:243, 1947.

[11] R. Georges, M. Freytes, D. Hurtmans, I. Kleiner, J. Vander Auwera, and M. Herman. Jet-cooled and room temperature FTIR spectra of the dimer of formic acid in the gas phase. *Chem. Phys.*, 305:187–196, 2004.

[12] Anders Engdahl and Bengt Nelander. Water in krypton matrices. *J. Mol. Struct*, 193:101–109, 1989.

[13] M. Halupka and W. Sander. A simple method for the matrix isolation of monomeric and dimeric carboxylic acids. *Spectrochim. Acta, Part A*, 54:495–500, 1998.

[14] M. Gantenberg, M. Halupka, and W. Sander. Dimerization of formic acid - an example of a noncovalent reaction mechanism. *Chem. Eur. J.*, 6:1865–1869, 2000.

[15] Fumiyuki Ito. Infrared spectra of $(HCOOH)_2$ and $(DCOOH)_2$ in rare gas matrices: A comparative study with gas phase spectra. *J. Chem. Phys.*, 128:114310, 2008.

[16] Y. Liu, M. Weimann, and M. A. Suhm. Extension of panoramic cluster jet spectroscopy into the far infrared: Low frequency modes of methanol and water clusters. *Phys. Chem. Chem. Phys.*, 6:3315–3319, 2004.

[17] Jens Dreyer. Hydrogen-bonded acetic acid dimers: Anharmonic coupling and linear infrared spectra studied with density-functional theory. *J. Chem. Phys.*, 122:184306, 2005.

[18] Tatsuo Miyazawa and Kenneth S. Pitzer. Internal rotation and infrared spectra of formic acid monomer and normal coordinate treatment of out-of-plane vibrations of monomer, dimer and polymer. *J. Chem. Phys.*, 30:1076–1086, 1959.

[19] James B. Togeas. Acetic acid vapor: 2. A statistical mechanical critique of vapor density experiments. *J. Phys. Chem. A*, 109:5438–5444, 2005.

[20] Z. Xue. Einflüsse von Konstitution und Konformation auf die O–H-Streckschwingungen gesättigter einwertiger C_5-Alkohole. Master's thesis, Universität Göttingen, 2007.

[21] M. Nedić. "Der kleinste Schnaps der Welt" – Experimentelle und theoretische Untersuchungen zu kleinen Alkohol-Wasser-Clustern. Hausarbeit im Rahmen der Ersten Staatsprüfung für das Lehramt an Gymnasien, Universität Göttingen 2008.

[22] M. Nedić, T. N. Wassermann, Z. Xue, P. Zielke, M. A. Suhm. Raman spectroscopic evidence for the most stable water/ethanol dimer and for the negative mixing energy in cold water/ethanol trimers. *Phys. Chem. Chem. Phys.*, 10:5953–5956, 2008.

[23] Franz Kollipost. Spektroskopische Untersuchungen zur Schwerflüchtigkeit von Lactonen und cyclischen Carbonaten. Master's thesis, U. Göttingen, 2009.

[24] Susanne Hesse and Martin A. Suhm. On the low volatility of cyclic esters: an infrared spectroscopy comparison between dimers of γ-butyrolactone and methyl propionate. *Phys. Chem. Chem. Phys.*, 11:11157–11170, 2009.

[25] Susanne Hesse. *Schwache Wechselwirkungen zwischen organischen Molekülen: Strukturelle Vielfalt und ihre schwingungsspektroskopischen Auswirkungen.* PhD thesis, Universität Göttingen, 2009.

[26] Tobias N. Wassermann, Jonas Thelemann, Philipp Zielke, and Martin A. Suhm. The stiffness of a fully stretched polyethylene chain: A Raman jet spectroscopy extrapolation. *J. Chem. Phys.*, 131:161108, 2009.

[27] B. Michielsen, J. J. J. Dom, B. J. van der Veken, S. Hesse, Z. Xue, M. A. Suhm, and W. A. Herrebout. The complexes of halothane with benzene: the temperature dependent direction of the complexation shift of the aliphatic C–H stretching. *Phys. Chem. Chem. Phys.*, 12:14034–14044, 2010.

[28] P. W. Atkins. *Physikalische Chemie.* Wiley-VCH, Weinheim, 3. Auflage, 2001.

[29] A. Smekal. Zur Quantentheorie der Dispersion. *Naturwissensch.*, 43:873–875, 1923.

[30] Rajinder Singh. C. V. Raman and the discovery of the Raman effect. *Physics in Perspective*, 4:399–420, 2002.

[31] G. Placzek. Rayleigh-Streuung und Raman-Effekt. In E. Marx, editor, *Quantenmechanik der Materie und Strahlung.* Akademische Verlagsgesellschaft m.b.H., Leipzig, 2. edition, 1934. Vol. VI, *Handbuch der Radiologie*, Kapitel 3, S. 205–374.

[32] D. A. Long. *The Raman Effect: A Unified Treatment of the Theory of Raman Scattering by Molecules.* John Wiley & Sons, Ltd, 2002.

[33] Philipp Zielke. *Ramanstreuung am Überschallstrahl: Wasserstoffbrückendynamik aus neuer Perspektive.* PhD thesis, U. Göttingen, 2007.

[34] C. L. Stevenson and T. Vo-Dinh. *Signal Expressions in Raman Spectroscopy.* Modern techniques in Raman spectroscopy. John Wiley & Sons, 1 edition, 1996.

[35] M. J. Frisch, G. W. Trucks, H. B. Schlegel, G. E. Scuseria, M. A. Robb, J. R. Cheeseman, J. A. Montgomery, Jr., T. Vreven, K. N. Kudin, J. C. Burant, J. M. Millam, S. S. Iyengar, J. Tomasi, V. Barone, B. Mennucci, M. Cossi, G. Scalmani, N. Rega, G. A. Petersson, H. Nakatsuji, M. Hada, M. Ehara, K. Toyota, R. Fukuda, J. Hasegawa, M. Ishida, T. Nakajima, Y. Honda, O. Kitao, H. Nakai, M. Klene, X. Li, J. E. Knox, H. P. Hratchian, J. B. Cross, C. Adamo, J. Jaramillo, R. Gomperts, R. E. Stratmann, O. Yazyev, A. J. Austin, R. Cammi, C. Pomelli, J. W. Ochterski, P. Y. Ayala, K. Morokuma, G. A. Voth, P. Salvador, J. J. Dannenberg, V. G. Zakrzewski, S. Dapprich, A. D. Daniels, M. C. Strain, O. Farkas, D. K. Malick, A. D. Rabuck, K. Raghavachari, J. B. Foresman, J. V. Ortiz, Q. Cui, A. G. Baboul, S. Clifford, J. Cioslowski, B. B. Stefanov, G. Liu, A. Liashenko, P. Piskorz, I. Komaromi, R. L. Martin, D. J. Fox, T. Keith, M. A. Al-Laham, C. Y. Peng, A. Nanayakkara, M. Challacombe, P. M. W. Gill, B. Johnson, W. Chen, M. W. Wong, C. Gonzalez, and J. A. Pople. Gaussian03, Revision B.04. Gaussian Inc., Pittsburgh PA, 2003.

[36] G. Keresztury. Raman Spectroscopy: Theory. In J. M. Chalmers, P. R. Griffiths, editor, *Handbook of Vibrational Spectroscopy*, volume 1: Theory and Instrumentation, chapter Introduction to the Theory and Practice of Vibrational Spectroscopy, pages 71–87. John Wiley & Sons Ltd., Chichester, 2002.

[37] N. Borho. *Chirale Erkennung in Molekülclustern: Maßgeschneiderte Aggregation von α-Hydroxyestern*. PhD thesis, Universität Göttingen, 2004.

[38] T. Häber. *Ragout-Jet-FTIR-Spektroskopie, Eine neuartige Methode zur Untersuchung der Dynamik von Oligomeren und nanometer-großen Molekülclustern*. PhD thesis, U. Göttingen, 2000.

[39] T. N. Wassermann. *Umgebungseinflüsse auf die C–C- und C–O-Torsionsdynamik in Molekülen und Molekülaggregaten: Schwingungsspektroskopie bei tiefen Temperaturen*. PhD thesis, Universität Göttingen, 2009.

[40] M. J. Frisch, G. W. Trucks, H. B. Schlegel, G. E. Scuseria, M. A. Robb, J. R. Cheeseman, G. Scalmani, V. Barone, B. Mennucci, G. A. Petersson, H. Nakatsuji, M. Caricato, X. Li, H. P. Hratchian, A. F. Izmaylov, J. Bloino, G. Zheng, J. L. Sonnenberg, M. Hada, M. Ehara, K. Toyota, R. Fukuda, J. Hasegawa, M. Ishida, T. Nakajima, Y. Honda, O. Kitao, H. Nakai, T. Vreven, J. A. Montgomery, Jr., J. E. Peralta, F. Ogliaro, M. Bearpark, J. J. Heyd, E. Brothers, K. N. Kudin, V. N. Staroverov, R. Kobayashi, J. Normand, K. Raghavachari, A. Rendell, J. C. Burant, S. S. Iyengar, J. Tomasi, M. Cossi, N. Rega, J. M. Millam, M. Klene, J. E. Knox, J. B. Cross, V. Bakken, C. Adamo, J. Jaramillo, R. Gomperts, R. E. Stratmann, O. Yazyev, A. J. Austin, R. Cammi, C. Pomelli, J. W. Ochterski, R. L. Martin, K. Morokuma, V. G. Zakrzewski, G. A. Voth, P. Salvador, J. J. Dannenberg, S. Dapprich, A. D. Daniels, O. Farkas, J. B. Foresman, J. V. Ortiz, J. Cioslowski, and D. J. Fox. **Gaussian 09, Revision A.02.** Gaussian, Inc., Wallingford CT, 2009.

[41] James B. Foresman and Æleen Frisch. *Exploring Chemistry with Electronic Structure Methods.* Gaussian, Inc. Pittsburgh, PA, 1996.

[42] Corinna Emmeluth. Schwingungsdynamik von Carbonsäureaggregaten. Master's thesis, U. Göttingen, 2001.

[43] Ilhan Yavuz and Carl Trindle. Structure, binding energies, and IR-spectral fingerprinting of formic acid dimers. *J. Chem. Theory Comput.*, 4:533–541, 2008.

[44] IUPAC Compendium of Chemical Terminology, Electronic version. http://goldbook.iupac.org/D01526.htm.

[45] C. A. Rice, N. Borho, and M. A. Suhm. Dimerization of pyrazole in slit jet expansions. *Z. Phys. Chem.*, 219:379–388, 2005.

[46] Nicole Borho, Martin A. Suhm, Katia Le Barbu-Debus, and Anne Zehnacker. Intra- vs. intermolecular hydrogen bonding: Dimers of alpha-hydroxyesters with methanol. *Phys. Chem. Chem. Phys.*, 8:4449–4460, 2006.

[47] Corinna Emmeluth, Volker Dyczmons, and Martin A. Suhm. Tuning the hydrogen bond donor/acceptor isomerism in jet-cooled mixed dimers of aliphatic alcohols. *J. Phys. Chem. A*, 110:2906–2915, 2006.

[48] T. Scharge, T. Häber, M. A. Suhm. Quantitative Chirality Synchronization in Trifluoroethanol Dimers. *Phys. Chem. Chem. Phys.*, 8:4664–4667, 2006.

[49] P. Zielke and M. A. Suhm. Concerted proton motion in hydrogen-bonded trimers: A spontaneous Raman scattering perspective. *Phys. Chem. Chem. Phys.*, 8:2826–2830, 2006.

[50] R. Wugt Larsen, Philipp Zielke, and Martin A. Suhm. Hydrogen-bonded OH stretching modes of methanol clusters: A combined IR and Raman isotopomer study. *J. Chem. Phys.*, 126:194307, 2007.

[51] M. Suhm. Spontane Ramanstreuung zur Charakterisierung der konzertierten Protonendynamik in isolierten Wasserstoffbrückenaggregaten. http://www.dfg.de/jahresbericht/Wce0a85735b151.htm, abgerufen am 28.05.2009.

[52] Tobias N. Wassermann, Philipp Zielke, Juhyon J. Lee, Christine Cézard, and Martin A. Suhm. Structural preferences, argon nanocoating, and dimerization of *n*-alkanols as revealed by OH stretching spectroscopy in supersonic jets. *J. Phys. Chem. A*, 111:7437–7448, 2007.

[53] Juhyon J. Lee, Sebastian Höfener, Wim Klopper, Tobias N. Wassermann, and Martin A. Suhm. Origin of the argon nanocoating shift in the OH stretching fundamental of *n*-propanol: A combined experimental and quantum chemical study. *J. Phys. Chem. C*, 113:10929–10938, 2009.

[54] Z. Xue and M. A. Suhm. Adding weight to a molecular recognition unit: The low-frequency modes of carboxylic acid dimers. *Mol. Phys.*, 108:2279–2288, 2010.

[55] Moritz Gadermann, Daniel Vollmar, and Ruth Signorell. Infrared spectroscopy of acetic acid and formic acid aerosols: pure and compound acid/ice particles. *Phys. Chem. Chem. Phys.*, 9:4535–4544, 2007.

[56] G. W. A. Kahlbaum. Studien über Dampfspannkraftsmessungen. *Z. Phys. Chem.*, 13:14–55, 1894.

[57] D. Ambrose. Thermodynamic properties of organic oxygen compounds XLIX. The vapour pressure of solid acetic acid. *J. Chem. Thermodyn.*, 11:183–185, 1979.

[58] R. A. McDonald, S. A. Shrader, and D. R. Stull. Vapor pressures and freezing points of 30 organics. *J. Chem. Eng. Data.*, 4:311–313, 1959.

[59] O. C. Bridgeman and E. W. Aldrich. Vapor pressure tables for water. *J. Heat Transfer*, 86:279–286, 1964.

[60] Bhajan S. Lark, Tarlok S. Banipal, and Surjit Singh. Excess gibbs energy for binary mixtures containing carboxylic acids. 3. excess gibbs energy for isobutyric acid and trimethylacetic acid + cyclohexane and + *n*-heptane. *J. Chem. Eng. Data.*, 32:402–406, 1987.

[61] Gautam R. Desiraju and Thomas Steiner. *The Weak Hydrogen Bond in Structural Chemistry and Biology.* Oxford University Press, Oxford, 1999.

[62] James T. Hynes, Judith P. Klinman, Hans-Heinrich Limbach, and Richard L. Schowen, editors. *Hydrogen-Transfer Reactions, Volume 1.* Wiley-VCH, 2007.

[63] Roman M. Balabin. Polar (acyclic) isomer of formic acid dimer: Gas-phase Raman spectroscopy study and thermodynamic parameters. *J. Phys. Chem. A*, 113:4910–4918, 2009.

[64] Mika Pettersson, Jan Lundell, Leonid Khriachtchev, and Markku Räsänen. IR spectrum of the other rotamer of formic acid, *cis*-HCOOH. *J. Am. Chem. Soc.*, 119:11715–11716, 1997.

[65] E. Bjarnov and W. M. Hocking. The structure of the other rotamer of formic acid, *cis*-HCOOH. *Z. Naturforsch.*, 33A:610, 1978.

[66] Kseniya Marushkevich, Leonid Khriachtchev, Jan Lundell, and Markku Räsänen. *cis–trans* Formic acid dimer: Experimental observation and improved stability against proton tunneling. *J. Am. Chem. Soc.*, 128:12060–12061, 2006.

[67] John E. Bertie, Kirk H. Michaelian, Hans H. Eysel, and Darcy Hager. The Raman-active O–H and O–D stretching vibrations and Raman spectra of gaseous formic acid-d_1 and -OD. *J. Chem. Phys.*, 85:4779–4789, 1986.

[68] Ermelinda M. S. Maçôas, Jan Lundell, Mika Pettersson, Leonid Khriachtchev, Rui Fausto, and Markku Räsänen. Vibrational spectroscopy of *cis*- and *trans*-formic acid in solid argon. *J. Mol. Spectrosc.*, 219:70–80, 2003.

[69] Adriana Olbert-Majkut, Jussi Ahokas, Jan Lundell, and Mika Pettersson. Raman spectroscopy of formic acid and its dimers isolated in low temperature argon matrices. *Chem. Phys. Lett.*, 468:176–183, 2009.

[70] Roger C. Millikan and Kenneth S. Pitzer. Infrared spectra and vibrational assignment of monomeric formic acid. *J. Chem. Phys.*, 27:1305, 1957.

[71] I. C. Hisatsune and J. Heicklen. Infrared spectrum of formyl chloride. *Can. J. Spectrosc.*, 18:135, 1973.

[72] Richard L. Redington and Kenneth C. Lin. On the OH stretching and the low-frequency vibrations of carboxylic acid cyclic dimers. *J. Chem. Phys.*, 54:4111, 1971.

[73] Oleg I. Baskakov, Igor A. Markov, Eugen A. Alekseev, Roman A. Motiyenko, Jarmo Lohilahti, Veli-Matti Horneman, Brenda P. Winnewisser, Ivan R. Medvedev, and Frank C. De Lucia. Simultaneous analysis of rovibrational and rotational data for the 4^1, 5^1, 6^1, 7^2, 8^1, 7^19^1 and 9^2 states of HCOOH. *J. Mol. Spectrosc.*, 795:54–77, 2006.

[74] Jean-Claude Deroche, Jyrki Kauppinen, and Esko Kyrö. ν_7 and ν_9 bands of formic acid near 16 μm. *J. Mol. Spectrosc.*, 78:379–394, 1979.

[75] R. C. Millikan and K. S. Pitzer. The infrared spectra of dimeric and crystalline formic acid. *J. Am. Chem. Soc.*, 80:3515–3521, 1958.

[76] Frank Madeja, Andreas Hecker, Simon Ebbinghaus, and Martina Havenith. High resolution spectroscopy of the ν_3 band of DCOOD. *Spectrochim. Acta Part A*, 59:1773–1782, 2003.

[77] Frank Madeja, Andreas Hecker, Simon Ebbinghaus, and Martina Havenith. High resolution spectroscopy of the ν_3 band of the van der Waals complex Ar-DCOOH. *Mol. Phys.*, 101:1511–1515, 2003.

[78] A. Almenningen, O. Bastiansen, and Tove Motzfeldt. A reinvestigation of the structure of monomer and dimer formic acid by gas electron diffraction technique. *Acta Chem. Scand.*, 23:2848–2864, 1969.

[79] Martina Havenith. Coherent proton tunneling in hydrogen bonds of isolated molecules. In J. T. Hynes, J. P. Klinman, H.-H. Limbach, and R. L. Schowen, editors, *Hydrogen-Transfer Reactions Vol. 1*, chapter 2, pages 33–51. Wiley-VCH, 2007.

[80] F. Carnovale, M. K. Livett, and J. B. Peel. A photoelectron spectroscopic study of the formic acid dimer. *J. Chem. Phys.*, 71:255–258, 1979.

[81] A. Winkler, J. B. Mehl, and P. Hess. Chemical relaxation of H bonds in formic acid vapor studied by resonant photoacoustic spectroscopy. *J. Chem. Phys.*, 100:2717–2727, 1994.

[82] M. Hippler. Proton relaxation and intermolecular structure of liquid formic acid: a nuclear magnetic resonance study. *Phys. Chem. Chem. Phys.*, 4:1457–1463, 2002.

[83] Y. Maréchal. IR spectra of carboxylic acids in the gas phase: A quantitative reinvestigation. *J. Chem. Phys.*, 87:6344–6353, 1987.

[84] V. V. Matylitsky, C. Riehn, M. F. Gelin, and B. Brutschy. The formic acid dimer $(HCOOH)_2$ probed by time-resolved structure selective spectroscopy. *J. Chem. Phys.*, 119:10553–10562, 2003.

[85] A. Gutberlet, G. W. Schwaab, and M. Havenith. High resolution IR spectroscopy of the carbonyl stretch of $(DCOOD)_2$. *Chem. Phys.*, 343:158–167, 2008.

[86] Yan-Tyng Chang, Yukio Yamaguchi, William H. Miller, and Henry F. Schaefer III. An analysis of the infrared and Raman spectra of the formic acid dimer $(HCOOH)_2$. *J. Am. Chem. Soc.*, 109:7245–7253, 1987.

[87] Isao Yokoyama, Yoshihisa Miwa, and Katsunosuke Machida. Simulation of Raman spectra of formic acid monomer and dimer in the gaseous state by an extended molecular mechanics method. *J. Phys. Chem.*, 95:9740–9746, 1991.

[88] Gina M. Florio, Timothy S. Zwier, Evgeniy M. Myshakin, Kenneth D. Jordan, and Edwin L. Sibert III. Theoretical modeling of the OH stretch infrared spectrum of carboxylic acid dimers based on first-principles anharmonic couplings. *J. Chem. Phys.*, 118:1735–1746, 2003.

[89] Christofer S. Tautermann, Andreas F. Voegele, and Klaus R. Liedl. The ground-state tunneling splitting of various carboxylic acid dimers. *J. Chem. Phys.*, 120:631–637, 2004.

[90] George L. Barnes, Shane M. Squires, and Edwin L. Sibert III. Symmetric double proton tunneling in formic acid dimer: A diabatic basis approach. *J. Phys. Chem. B*, 112:595–603, 2008.

[91] George L. Barnes and Edwin L. Sibert III. An equilibrium focused approach to calculating the Raman spectrum of the symmetric OH stretch in formic acid dimer. *J. Mol. Spectrosc.*, 249:78–85, 2008.

[92] David Luckhaus. Concerted hydrogen exchange tunneling in formic acid dimer. *J. Phys. Chem. A*, 110:3151–3158, 2006.

[93] I. Matanović, N. Došlić, and O. Kühn. Ground and asymmetric CO-stretch excited state tunneling splittings in the formic acid dimer. *J. Chem. Phys.*, 127:014309, 2007.

[94] George L. Barnes and Edwin L. Sibert III. The effects of asymmetric motions on the tunneling splittings in formic acid dimer. *J. Chem. Phys.*, 129:164317, 2008.

[95] Fumiyuki Ito. Jet-cooled infrared spectra of the formic acid dimer by cavity ring-down spectroscopy: Observation of the C–O stretching region and vibrational analysis of the Fermi-triad system. *Chem. Phys. Lett.*, 447:202–207, 2007.

[96] Richard J. Saykally. Far-infrared laser spectroscopy of van der Waals bonds: A powerful new probe of intermolecular forces. *Acc. Chem. Res.*, 22:295–300, 1989.

[97] Katharina von Puttkamer, Martin Quack, and Martin A. Suhm. Infrared spectrum and dynamics of the hydrogen bonded dimer (HF)$_2$. *Infrared Phys.*, 29:535–539, 1989.

[98] Y. Heidi Yoon, Michael L. Hause, Amanda S. Case, and F. Fleming Crim. Vibrational action spectroscopy of the C–H and C–D stretches in partially deuterated formic acid dimer. *J. Chem. Phys.*, 128:084305, 2008.

[99] Chayan K. Nandi, Montu K. Hazra, and Tapas Chakraborty. Vibrational coupling in carboxylic acid dimers. *J. Chem. Phys.*, 123:124310, 2005.

[100] David T. Anderson, Scott Davis, and David J. Nesbitt. Hydrogen bond spectroscopy in the near infrared: Out-of-plane torsion and antigeared bend combination bands in $(HF)_2$. *J. Chem. Phys.*, 105:4488–4503, 1996.

[101] Thomas Häber, Ulrich Schmitt, Corinna Emmeluth, and Martin A. Suhm. Ragout-jet FTIR spectroscopy of cluster isomerism and cluster dynamics: from carboxylic acid dimers to N_2O nanoparticles. *Faraday Discuss.*, 118:331–359, 2001. +contributions to the discussion on pp 53, 119, 174–175, 179–180, 304–309, 361–363, 367–370.

[102] Corinna Emmeluth, Martin A. Suhm, and David Luckhaus. A monomers-in-dimers model for carboxylic acid dimers. *J. Chem. Phys.*, 118:2242–2255, 2003.

[103] Frank Madeja, Martina Havenith, Klaas Nauta, Roger E. Miller, Jana Chocholousova, and Pavel Hobza. Polar isomer of formic acid dimers formed in helium droplets. *J. Chem. Phys.*, 120:10554–10560, 2004.

[104] Steven T. Shipman, Pamela C. Douglass, Hyun S. Yoo, Charlotte E. Hinkle, Ellen L. Mierzejewski, and Brooks H. Pate. Vibrational dynamics of carboxylic acid dimers in gas and dilute solution. *Phys. Chem. Chem. Phys.*, 9:4572–4586, 2007.

[105] G. L. Carlson, R. E. Witkowski, and W. G. Fateley. Far infrared spectra of dimeric and crystalline formic and acetic acids. *Spectrochim. Acta*, 22:1117–1123, 1966.

[106] R. J. Jakobsen, J. Mikawa, and J. W. Brasch. Far infrared studies of hydrogen bonding in carboxylic acids—I Formic and acetic acids. *Spectrochim. Acta A*, 23:2199–2209, 1967.

[107] Markus Ortlieb and Martina Havenith. Proton transfer in $(HCOOH)_2$: An IR high-resolution spectroscopic study of the antisymmetric C–O stretch. *J. Phys. Chem. A*, 111:7355–7363, 2007.

[108] Fumiyuki Ito and Taisuke Nakanaga. A jet-cooled infrared spectra of the formic acid dimer by cavity ring-down spectroscopy. *Chem. Phys. Lett.*, 318:571–577, 2000.

[109] Fumiyuki Ito and Taisuke Nakanaga. Jet-cooled infrared spectra of the formic acid dimer by cavity ring-down spectroscopy: observation of the O–H stretching region. *Chem. Phys.*, 277:163–169, 2002.

[110] U. Merker, P. Engels, F. Madeja, M. Havenith, and W. Urban. High-resolution CO-laser sideband spectrometer for molecular-beam optothermal spectroscopy in the 5-6.6μm wavelength region. *Rev. Sci. Instr.*, 70:1933–1938, 1999.

[111] F. Madeja and M. Havenith. High resolution spectroscopy of carboxylic acid in the gas phase: Observation of proton transfer in $(DCOOH)_2$. *J. Chem. Phys.*, 117:7162–7168, 2002.

[112] Kozo Hirota and Yasuo Nakai. Far infrared spectrum of gaseous formic acid. *Bull. Chem. Soc. Jpn.*, 32:769–771, 1959.

[113] D. Clague and A. Novak. Far infrared spectra of homogeneous and heterogeneous dimers of some carboxylic acids. *J. Mol. Struct.*, 5:149–152, 1970.

[114] G. V. Mil'nikov, O. Kühn, and H. Nakamura. Ground-state and vibrationally assisted tunneling in the formic acid dimer. *J. Chem. Phys.*, 123:074308, 2005.

[115] Vincenzo Barone. Anharmonic vibrational properties by a fully automated second-order perturbative approach. *J. Chem. Phys.*, 122:014108, 2005.

[116] L. G. Bonner and J. S. Kirby-Smith. The Raman spectrum of formic acid vapor. *Phys. Rev.*, 57:1078, 1940.

[117] C. Emmeluth and M. A. Suhm. A chemical approach towards the spectroscopy of carboxylic acid dimer isomerism. *Phys. Chem. Chem. Phys.*, 5:3094–3099, 2003.

[118] L. Martinache, W. Kresa, M. Wegener, U. Vonmont, and A. Bauder. Microwave spectra and partial substitution structure of carboxylic acid bimolecules. *Chem. Phys.*, 148:129–140, 1990.

[119] Stefan Schweiger and Guntram Rauhut. Double proton transfer reactions with plateau-like transition state regions: Pyrazole-trifluoroacetic acid clusters. *J. Phys. Chem. A*, 110:2816–2820, 2006.

[120] K. L. Goh, P. P. Ong, and T. L. Tan. The ν_3 band of DCOOH. *Spectrochimica Acta Part A*, 55:2601–2614, 1999.

[121] Zorka Smedarchina, Antonio Fernández-Ramos, and Willem Siebrand. Tunneling dynamics of double proton transfer in formic acid and benzoic acid dimers. *J. Chem. Phys.*, 122:134309, 2005.

[122] David Luckhaus. Hydrogen exchange in formic acid dimer: tunnelling above the barrier. *Phys. Chem. Chem. Phys.*, 12:8357–8361, 2010.

[123] Ö. Birer and M. Havenith. High resolution infrared spectroscopy of formic acid dimer. *Annu. Rev. Phys. Chem.*, 60:263–275, 2009.

[124] Y. Nakai and K. Hirota. Far-infrared spectra of gaseous formic acid and acetic acid. *J. Chem. Soc. Japan*, 81:881, 1960.

[125] D. J. Frurip, L. A. Curtiss, and M. Blander. Vapor phase association in acetic and trifluoroacetic acids. Thermal conductivity measurements and molecular orbital calculations. *J. Am. Chem. Soc*, 102:2610–2616, 1980.

[126] H. R. Zelsmann, Z. Mielke, and Y. Marechal. Far IR spectra of acetic acids in the gas phase. A reinvestigation of the intermonomer vibrations. *J. Mol. Struct.*, 237:273–283, 1990.

[127] Robert Gaufrès, Jacques Maillols, and Vlado Tabacik. Composition of a gaseous mixture in chemical equilibrium as studied by Raman spectroscopy: Gas phase acetic acid. *J. Raman Spectrosc.*, 11:442–448, 1981.

[128] John E. Bertie and Kirk H. Michaelian. The Raman spectrum of gaseous acetic acid at 21°C. *J. Chem. Phys.*, 77:5267–5271, 1982.

[129] Jens Dreyer. Density Functional Theory Simulations of Two-Dimensional Infrared Spectra for Hydrogen-Bonded Acetic Acid Dimers. *J. Quant. Chem.*, 104:782–793, 2005.

[130] Mohamed El-Amine Benmalti, Paul Blaise, H. T. Flakus, and Olivier Henri-Rousseau. Theoretical interpretation of the infrared lineshape of liquid and gaseous acetic acid. *Chem. Phys.*, 320:267–274, 2005.

[131] Takakazu Nakabayashi, Kentaroh Kosugi, and Nobuyuki Nishi. Liquid structure of acetic acid studied by Raman spectroscopy and ab initio molecular orbital calculations. *J. Phys. Chem. A*, 103:8595–8603, 1999.

[132] M. Haurie and A. Novak. Vibrational spectra of the CH_3COOH, CH_3COOD, CD_3COOH, and CD_3COOD molecules. II. Infrared and Raman spectra of the dimers. *J. Chim. Phys. Phys. Phys.-Chim. Biol.*, 62:146–157, 1965.

[133] Martti Ovaska. Polarized infrared spectra of hydrogen-bonded systems by the stretched-polymer method. 2. transition moment directions of the vibrations of normal and deuterated acetic acid dimers. *J. Phys. Chem.*, 88:5981–5986, 1984.

[134] O. Faurskov Nielsen and P.-A. Lund. Intermolecular Raman active vibrations of hydrogen bonded acetic acid dimers in the liquid state. *J. Chem. Phys.*, 78:652–655, 1983.

[135] Karsten Heyne, Nils Huse, Erik T. J. Nibbering, and Thomas Elsaesser. Ultrafast coherent nuclear motions of hydrogen bonded carboxylic acid dimers. *Chem. Phys. Lett.*, 369:591–596, 2003.

[136] Robert E. Jones and David H. Templeton. The crystal structure of acetic acid. *Acta Cryst.*, 5:484–487, 1958.

[137] Alfred E. Beylich. Struktur von Ueberschall-Freistrahlen aus Schlitzblenden. *Z. Flugwiss. Weltraumforsch.*, 3:48–58, 1979.

[138] C. Emmeluth, T. Häber, and M. A. Suhm. General discussion. *Faraday Discuss.*, 118:361–371, 2001.

[139] Stefan Grimme, Jens Antony, Tobias Schwabe, and Christian Mück-Lichtenfeld. Density functional theory with dispersion corrections for supramolecular structures, aggregates, and complexes of (bio)organic molecules. *Org. Biomol. Chem.*, 5:741–758, 2007.

[140] A. Winkler and P. Hess. Study of the energetics and dynamics of hydrogen bond formation in aliphatic carboxylic acid vapors by resonant photoacoustic spectroscopy. *J. Am. Chem. Soc.*, 116:9233–9240, 1994.

[141] W. Longueville and Fontaine. Diffusion Raman des Acides Pivalique Hydrogéné et Déutérie en Phase Liquide et en Solution. *J. Raman Spectrosc.*, 7:238–243, 1978.

[142] W. Longueville, H. Fontaine, and G. Vergoten. Spectre de Vibration et Vibrations Normales de l'Acide Pivalique. *J. Raman Spectrosc.*, 13:213–222, 1982.

[143] M. Chhiba, P. Derreumaux, and G. Vergoten. The use of the SPASIBA spectroscopic potential for reproducing the structures and vibrational frequencies of a sries of acids: acetic acid, pivalic acid, succinic acid, adipic acid and L-glutamic acid. *J. Mol. Spectrosc.*, 317:171–184, 1994.

[144] Krešimir Furić and Vesna Volovšek. Water ice at low temperatures and pressures: New Raman results. *J. Mol. Spectrosc.*, 976:174–180, 2010.

[145] C. Camy-Peyret, J. M. Flaud, G. Guelachvili, and C. Amiot. High resolution Fourier transform spectrum of water between 2930 and 4255 cm^{-1}. *Mol. Phys.*, 26:825–855, 1973.

[146] N. Papineau, C. Camy-Peyret, J. M. Flaud, and G. Guelachvili. The $2\nu_2$ and ν_1 bands of water-D_1. *J. Mol. Spectrosc.*, 92:451–468, 1982.

[147] G. Avila, J. M. Fernández, B. Maté, G. Tejeda, and S. Montero. Rovibrational Raman Cross Sections of Water Vapor in the OH Stretching Region. *J. Mol. Spectrosc.*, 196:77–92, 1999.

[148] G. Avila, G. Tejeda, J. M. Fernández, and S. Montero. The rotational Raman spectra and cross sections of H_2O, D_2O, and HDO. *J. Mol. Spectrosc.*, 220:259–275, 2003.

[149] G. Avila, G. Tejeda, J. M. Fernández, and S. Montero. The rotational Raman spectra and cross sections of the ν_2 band of H_2O, D_2O, and HDO. *J. Mol. Spectrosc.*, 223:166–180, 2004.

[150] G. Avila, G. Tejeda, J. M. Fernández, and S. Montero. The rotational Raman spectra and cross sections of H_2O, D_2O, and HDO in the OH/OD stretching regions. *J. Mol. Spectrosc.*, 228:38–65, 2004.

[151] M. F. Vernon, D. J. Krajnovich, H. S. Kwok, J. M. Lisy, Y. R. Shen, and Y. T. Lee. Infrared vibrational predissociation spectroscopy of water clusters by the crossed laser-molecular beam technique. *J. Chem. Phys.*, 77:47, 1982.

[152] Ralph H. Page, Jeremy G. Frey, Y.-R. Shen, and Y. T. Lee. Infrared predissociation spectra of water dimer in a supersonic molecular beam. *Chem. Phys. Lett.*, 106:373–376, 1984.

[153] D. F. Coker, R. E. Miller, and R. O. Watts. The infrared predissociation spectra of water clusters. *J. Chem. Phys.*, 82:3554, 1985.

[154] Stefan Wuelfert, Daniel Herren, and Samuel Leutwyler. Supersonic jet CARS spectra of small water clusters. *J. Chem. Phys.*, 86:3751–3753, 1987.

[155] Anders Engdahl and Bengt Nelander. On the structure of the water trimer. A matrix isolation study. *J. Chem. Phys.*, 86:4831, 1987.

[156] Bengt Nelander. The intramolecular fundamentals of the water dimer. *J. Chem. Phys.*, 88:5254–5256, 1988.

[157] Ralf Fröchtenicht, Michael Kaloudis, Martin Koch, and Friedrich Huisken. Vibrational spectroscopy of small water complexes embedded in large liquid helium clusters. *J. Chem. Phys.*, 105:6128, 1996.

[158] Friedrich Huisken, Michael Kaloudis, and Axel Kulcke. Infrared spectroscopy of small size-selected water clusters. *J. Chem. Phys.*, 104:17–25, 1996.

[159] J. B. Paul, C. P. Collier, R. J. Saykally, J. J. Scherer, and A. O'Keefe. Direct measurement of water cluster concentrations by infrared cavity ringdown laser absorption spectroscopy. *J. Phys. Chem. A*, 101:5211–5214, 1997.

[160] J. B. Paul, R. A. Provencal, C. Chapo, A. Petterson, and R. J. Saykally. Infrared cavity ringdown spectroscopy of water clusters: O–D stretching bands. *J. Chem. Phys.*, 109:10201, 1998.

[161] Lisa M. Goss, Steven W. Sharpe, Thomas A. Blake, Veronica Vaida, and James W. Brault. Direct absorption spectroscopy of water clusters. *J. Phys. Chem. A*, 103:8620–8624, 1999.

[162] Frank N. Keutsch, Jeffery D. Cruzan, and Richard J. Saykally. The water trimer. *Chem. Rev.*, 103:2533–2577, 2003.

[163] K. Nauta and R. E. Miller. Formation of cyclic water hexamer in liquid helium: The smallest piece of ice. *Science*, 287:293–295, 2000.

[164] D. N. Nesbitt, T. Häber, and M. A. Suhm. General discussion. *Faraday Discuss.*, 118:295–314, 2001.

[165] Keiichi Ohno, Mari Okimura, Nobuyuki Akai, and Yukiteru Katsumoto. The effect of cooperative hydrogen bonding on the OH stretching-band shift for water clusters studied by matrix-isolation infrared spectroscopy and density functional theory. *Phys. Chem. Chem. Phys.*, 7:3005–3014, 2005.

[166] Justinas Ceponkus, Gunnar Karlström, and Bengt Nelander. Intermolecular vibrations of the water trimer, a matrix isolation study. *J. Phys. Chem. A*, 109:7859–7864, 2005.

[167] Joseph E. Fowler and Henry F. Schaefer III. Detailed study of the water trimer potential energy surface. *J. Am. Chem. Soc.*, 117:446–452, 1995.

[168] A. van der Avoird, E. H. T. Olthof, and P. E. S. Wormer. Tunneling dynamics, symmetry, and far-infrared spectrum of the rotating water trimer. I. Hamiltonian and qualitative model. *J. Chem. Phys.*, 105:8034–8050, 1996.

[169] Joon O. Jung and R. Benny Gerber. Vibrational wave functions and spectroscopy of $(H_2O)_n$, $n = 2,3,4,5$: Vibrational self-consistent field with correlation corrections. *J. Chem. Phys.*, 105:10332, 1996.

[170] G. C. Groenenboom, E. M. Mas, R. Bukowski, K. Szalewicz, P. E. S. Wormer, and A. van der Avoird. Water pair and three-body potential of spectroscopic quality from *ab initio* calculations. *Phys. Rev. Lett.*, 84:4072–4075, 2000.

[171] Sotiris S. Xantheas. Cooperativity and hydrogen bonding network in water clusters. *Chem. Phys.*, 258:225–231, 2000.

[172] Christian J. Burnham, Sotiris S. Xantheas, Mark A. Miller, Brian E. Applegate, and Roger E. Miller. The formation of cyclic water complexes by sequential ring insertion: Experiment and theory. *J. Chem. Phys.*, 117:1109, 2002.

[173] Meghan E. Dunn, Timothy M. Evans, Karl N. Kirschner, and George C. Shields. Prediction of accurate anharmonic experimental vibrational frequencies for water clusters, $(H_2O)_n$, $n = 2$-5. *J. Phys. Chem. A*, 110:303–309, 2006.

[174] Julie A. Anderson, Kelly Crager, Lisa Fedoroff, and Gregory S. Tschumper. Anchoring the potential energy surface of the cyclic water trimer. *J. Chem. Phys.*, 121:11023, 2004.

[175] Yimin Wang, Stuart Carter, Bastiaan J. Braams, and Joel M. Bowman. MULTIMODE quantum calculations of intramolecular vibrational energies of the water dimer and trimer using *ab initio*-based potential energy surfaces. *J. Chem. Phys.*, 128:071101, 2008.

[176] J. B. Paul, R. A. Provencal, C. Chapo, K. Roth, R. Casaes, and R. J. Saykally. Infrared cavity ringdown spectroscopy of the water clusters bending vibrations. *J. Phys. Chem. A*, 103:2972–2974, 1999.

[177] A. Moudens, R. Georges, M. Goubet, J. Makarewicz, S. E. Lokshtanov, and Vigasin A. A. Direct absorption spectroscopy of water clusters formed in a continuous slit nozzle expansion. *J. Chem. Phys.*, 131:204312, 2009.

Lebenslauf

Am 15.12.1981 wurde ich als älteres von zwei Kindern der Schneiderin Fengxiang Xue, geborene Wang, und des Autoingenieurs Guiman Xue in Qingdao, Shandong, V. R. China geboren. Dort verbrachte ich meine Kindheit und besuchte von 1987 bis 1993 die Grundschule, von 1993 bis 1996 die allgemeine Mittelschule und von 1996 bis 1999 die Oberstufe der allgemeinen Mittelschule. Im Juli 1999 nahm ich an der staatlichen einheitlichen Hochschulaufnahmeprüfung teil und wurde ins Fach angewandte Chemie der Fakultät für Chemie der Tongji-Universität in Shanghai aufgenommen. Im September 1999 immatrikulierte ich mich für den Bachelor-Studiengang Chemie an der Tongji-Universität. Danach ging ich im März 2003 nach Deutschland und besuchte vom April 2003 bis zum Juli 2003 den Deutschkurs für ausländische Studenten an der Georg-August-Universität Göttingen. Am 19.09.2003 nahm ich erfolgreich an der Deutschen Sprachprüfung für den Hochschulzugang ausländischer Studienbewerber (DSH) teil.

Im Oktober 2003 immatrikulierte ich mich im Diplom-Studiengang Chemie an der Georg-August-Universität Göttingen. Am 26.10.2005 legte ich meine Diplomvorprüfung ab. Von März 2007 bis September 2007 habe ich meine Diplomarbeit zum Thema „Einflüsse von Konstitution und Konformation auf die O–H-Streckschwingungen gesättigter einwertiger C_5-Alkohole" unter der Leitung von Prof. Dr. M. A. Suhm gemacht. Am 31.10.2007 bestand ich die Diplomprüfung.

Seit dem 01.11.2007 bin ich Mitglied des Graduiertenkollegs GRK 782 und beschäftigte ich mich in meiner Doktorarbeit mit dem Thema „Raman Spektroskopie von Carbonsäure- und Wasseraggregaten" (Raman spectroscopy of carboxylic acid and water aggregates) in der Arbeitsgruppe von Herrn Prof. Dr. M. A. Suhm am Institut für Physikalische Chemie der Georg-August-Universität Göttingen. Als Ergebnis dieser Tätigkeit entstand die vorliegende Arbeit.

Im Dezember 2004 habe ich Fei Zhou geheiratet, Nachwuchs ist unterwegs.